the ocean book

DK

Smithsonian

the ocean book

THE STORIES, SCIENCE, AND HISTORY OF OCEANS

Senior Editor Hugo Wilkinson	**Senior Art Editor** Duncan Turner
Editor Ian Fitzgerald	**Design Assistant** Noor Ali
US Editor Megan Douglass	**Picture Researcher** Laura Barwick
Senior DTP Designer Harish Aggarwal	**Senior Jackets Designer** Surabhi Wadhwa-Gandhi
Senior Production Editor Andy Hilliard	**Managing Art Editor** Michael Duffy
Senior Production Controller Meskerem Berhane	**Art Director** Maxine Pedliham
Managing Editor Angeles Gavira	**Design Director** Phil Ormerod
Publishing Director Georgina Dee	
Managing Director Liz Gough	

Produced for DK by Dynamo Limited
1 Cathedral Court, Southernhay East,
Exeter EX1 1AF
Editorial Partner Claire Lister
Design Partner Jeremy Marshall

DK Delhi
Senior Art Editor Anjali Sachar
Jacket Designer Juhi Sheth
Senior DTP Designer Harsh Aggarwal
Senior Jackets Coordinator Priyanka Sharma Saddi

First American Edition, 2025
Published in the United States by DK Publishing,
a division of Penguin Random House LLC
1745 Broadway, 20th Floor, New York, NY 10019

Copyright © 2025 Dorling Kindersley Limited
24 25 26 27 28 10 9 8 7 6 5 4 3 2 1

001–342553–May/2025

All rights reserved.

Without limiting the rights under the copyright reserved above, no part of this publication may be reproduced, stored in or introduced into a retrieval system, or transmitted, in any form, or by any means (electronic, mechanical, photocopying, recording, or otherwise), without the prior written permission of the copyright owner.

Published in Great Britain by Dorling Kindersley Limited

A catalog record for this book is available from the Library of Congress.

ISBN 978-0-5939-6362-3

DK books are available at special discounts when purchased in bulk for sales promotions, premiums, fund-raising, or educational use. For details, contact: DK Publishing Special Markets, 1745 Broadway, 20th Floor, New York, NY 10019 SpecialSales@dk.com

Printed and bound in India

www.dk.com

Contributors

Jason Hall-Spencer (consultant) is one of the world's leading experts investigating the major stressors affecting the health of our seas and the marine organisms impacted by climate change. He is a professor of Marine Biology at the Universities of Plymouth (UK) and Tsukuba (Japan).

Dorrik Stow is a geologist and oceanographer who has pioneered research on deep-sea sediments and paleocirculation including how these are linked with climate and past environmental change. He is an author of numerous books and scientific publications, has worked across the world, and is now Professor Emeritus at Heriot Watt University in Edinburgh.

Melissa Hobson is a writer who specialises in marine science, conservation, and sustainability. A keen scuba diver, she works with NGOs and organisations to help them achieve their communication goals regarding the ocean.

Anna Claybourne is a science writer who has a passion for nature and wildlife. She has written more than 300 books on many different subjects. An avid sea swimmer, camper, and wildlife-watcher, she also loves mythology and folklore, and writes about mermaids and sea monsters as well as countless real-life creatures.

Bess Manley has a zoology degree and spent more than a decade working on tropical coral reefs in Fiji and Australia. She has written and directed documentaries for Channel 4, Discovery Channel, and Nat Geo, as well as non-broadcast content for marine charities, NGOs, and government organizations.

Marianne Taylor is a life-long birder and wildlife enthusiast who has worked since 2007 as a freelance writer, editor, illustrator, and photographer. A former editor at the publisher Bloomsbury, she has written more than 30 books on natural history, in particular birds and insects, and regularly works in collaboration with the RSPB.

For Smithsonian National Museum of Natural History: Darrin Lunde, Collection Manager
For Smithsonian Enterprises:
Avery Naughton, Licensing Coordinator
Paige Towler, Editorial Lead
Jill Corcoran, Senior Director, Licensed Publishing
Brigid Ferraro, Vice President of New Business and Licensing

Carol LeBlanc, President

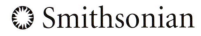

This trademark is owned by the Smithsonian Institution and is registered in the U.S. Patent and Trademark Office. Established in 1846, the Smithsonian Institution—the world's largest museum and research complex—includes 19 museums and galleries and the National Zoological Park. The total number of artifacts, works of art, and specimens in the Smithsonian's collections is estimated at 156 million, the bulk of which is contained in the National Museum of Natural History, which holds more than 126 million specimens and objects. The Smithsonian is a renowned research center, dedicated to public education, national service, and scholarship.

This book was made with Forest Stewardship Council™ certified paper – one small step in DK's commitment to a sustainable future.
Learn more at www.dk.com/uk/information/sustainability

contents

CHAPTER 1
How the Ocean Works

Ocean and Seas	12
Evolution and Geology of the Ocean	16
Seawater	20
Surface Currents	22
Connected Ocean	24
Ocean and Climate	26
Tides	28
Coastlines	30
Ocean Zones	32
Ocean Animals	34
Plants and Algae	36

CHAPTER 2
Ocean Edges

Cliffs	40
Puffins	42
Arches and Sea Stacks	44
Cormorants	48
Sea Caves	50
Gastropods	52
Rocky Shores	56
Seals	60
Coastal Fossils	62
Rock Pools	64
Sea Urchins	66
Crustaceans	68
Rocky Beaches	72
Yellow Horned Poppy	74
Coastal Lagoons	76
Flamingos	78

Chesil Beach	80
Tied Islands and Tombolos	82
Sandy Beaches	84
Sea Turtles	88
Australian Pelican	94
Coconut Palm	96
Lugworms	98
Marram Grass	99
Dunes	100
Seagrasses	102
Skeleton Coast	104
Sea Lions	106
Estuaries and Deltas	108
Dugong and Manatees	112
Fjords	114
Beluga Whale	116
Kelp Forests	118
Sea Otter	120
Salt Marshes and Mudflats	122
Waders	126
Mangroves	128
Saltwater Crocodiles	130
Desert Islands	132
Galápagos	136

CHAPTER 3
On and Above the Surface

Ocean Winds	140
Albatrosses	144
Tropical Cyclones	146
Waves	148
Tsunamis	154
Gulls	156
Sargasso Sea	158
By-the-Wind Sailor	160
Whirlpools	162
Sea Mists and Fogs	164
Terns	166

CHAPTER 4
Icy Oceans

Ice Shelves and Icebergs	170
Sea Ice	172
Polar Bear	174
Narwhal	178
Arctic Ocean	180
Walrus	182
Southern Ocean	184
Penguins	188
Leopard Seal	192

CHAPTER 5
Beneath the Waves

Sunlit Zone	196
Plankton	198
Seaweeds	202
Coral Reefs	204
Seahorses	208
Great Barrier Reef	210
Leafy Sea Dragon	212
Eels	214
Jellyfish	216
Dolphins	220
Orca	222
Whales	224
Atlantic Ocean	230
Echinoderms	234
Rocky Reefs	236
Sunken Cities	238
Sharks	240
Stingrays	244
Ocean Floor	248
Twilight Zone	250
Giant Oarfish	252
Octopuses	254
Pacific Ocean	258
Hammerhead Sharks	262
Sea Snakes	264
Billfish	268
Bivalves	270
Tuna	274
Squid	278
Indian Ocean	280
Dark Zone	284
Bioluminescence	286
Chimaeras	288
Sponges	290
Volcanoes and Seamounts	292
Hydrothermal Vents	294
Giant Tube Worm	296
Sea Pigs	297
Abyssal and Hadal Zones	298
Trenches	300

Glossary	304
Index	308
Acknowledgments	318

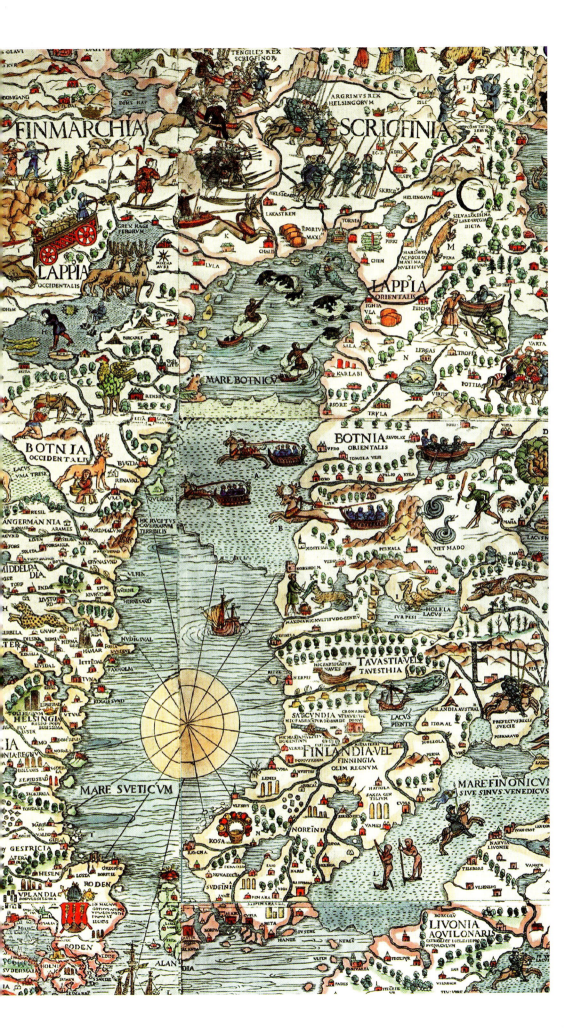

Mapping the seas

Published in 1539, the *Carta marina et descriptio septemtrionalium terrarum* (Marine map and description of the Northern lands) was the first known map to depict the Nordic countries, and their surrounding waters in detail. Sea monsters reflect the danger and mystery the ocean held for early explorers. Swedish cartographer and clergyman Olaf Månsson took 12 years to produce the enormous 49 in by 67 in (125 cm by 170 cm) map, drawing on ancient and contemporary sources as well as the accounts of sailors.

CHAPTER 1

How the Ocean Works

The ocean is a complex, ever-changing system that affects every aspect of life on Earth. It regulates the planet's temperature, weather, and atmosphere, and it supports a range of thriving ecosystems of immense diversity.

Ocean and Seas

With around 71 percent of its surface covered by water, Earth is often called the Blue Planet. The ocean and seas together hold 96.5 percent of the planet's water, with most of the rest locked up in ice or buried deep below the land surface.

> Around 80 **percent** of Earth's **oxygen** is produced by photosynthesizing **microscopic plankton**

What is an ocean?

An ocean is a large body of saltwater. Earth has one global ocean, divided up into five interconnected ones. The ocean fills deep depressions in Earth's surface, which have sediment-covered floors underpinned by a crust composed mainly of basalt—a dense, dark volcanic rock. New oceanic crust is continually being formed at mid-ocean ridges and consumed in submarine trenches (see pp.18–19). This means the size and shape of the ocean above is forever changing. Life first evolved in the ocean around four billion years ago, and today it supports a vast diversity of species, from microscopic plankton to the largest animal on the planet, the blue whale (*Balaenoptera musculus*). The ocean is also the driving force that regulates Earth's climate.

Five regions

Earth's five interconnected ocean basins are the Arctic, Atlantic, Indian, Pacific, and Southern (sometimes Antarctic). The last of these was only named in 1999 and recognized by the National Geographic Society in 2021. Where one ocean basin ends and another begins is not strictly defined. Together, they cover 81 percent of the southern hemisphere and 61 percent of the northern hemisphere.

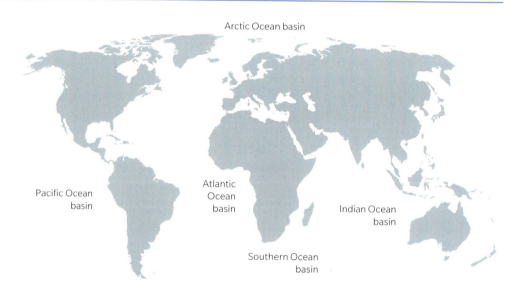

Name	Area	Average depth	Maximum depth	Oldest crust
Arctic Ocean basin	4,700,000 sq miles (12,173,000 sq km)	3,250 ft (990 m)	18,399 ft (Molloy Deep) (5,608 m)	65 million years
Atlantic Ocean basin	32,000,00 sq miles (82,000,000 sq km)	10,800 ft (3,300 m)	28,232 ft (Milwaukee Deep) (8,605 m)	175 million years
Indian Ocean basin	28,400,000 sq miles (73,600,000 sq km)	12,300 ft (3,740 m)	24,450 ft (Java Trench) (7,450 m)	140 million years
Pacific Ocean basin	64,000,000 sq miles (166,000,000 sq km)	14,050 ft (4,280 m)	35,840 ft (Challenger Deep) (10,924 m)	165 million years
Southern Ocean basin	13,500,000 sq miles (35,000,000 sq km)	10,700 ft (3,270 m)	23,738 ft (South Sandwich Trench) (7,235 m)	140 million years

OCEAN AND SEAS

Geography of the ocean floor

Deep beneath the ocean surface, the ocean floor is far from featureless. It has the longest mountain chains on the planet, cut through by deep chasms known as transform faults, which are slowly smoothed by sediment cover away from their axis. Where the ocean meets continents, gently sloping shelves soon plunge seaward down steeper slopes, sculpted into canyons and slide scars, to reach flat abyssal plains or mountainous drifts of mud. Ripples, dunes, and giant waves move across the sediment-covered surface, propelled by deep ocean current activity. Numerous seamounts and guyots—undersea mountains—pierce through the sediment cover, reaching into the dark, twilight, or sunlit ocean zones.

Surveying Silfra
The Mid-Atlantic Ridge runs along the Atlantic floor on its north-south axis. Where it emerges and traverses Iceland, its central deep rift is exposed. The Silfra Fissure, now an Icelandic lake, was formed in the rift by earthquake activity.

Mapping the ocean

Although Earth's entire surface has been mapped, only around 5 percent of the ocean has been explored in detail. Technology such as sonar (sound navigation and ranging), satellite imaging, and manned and unmanned submersibles are used to analyze and map the ocean floor.

Hull made from titanium alloy

Ocean floor explorer
This manned submersible from the French Research Institute of the Sea (IFREMER) was designed to operate at depths of up to 3¾ miles (6 km).

Extreme depths

The deepest ocean zones (see pp.298–299) feature cold temperatures, extreme pressure, and darkness. Without photosynthesis, food sources are scarce, but some animals survive at this depth. Hydrothermal vents (see pp.294–295) form unique deep-ocean ecosystems.

Life at the vent
The "Champagne" hydrothermal vent on the floor of the western Pacific is home to a large population of mussels and crabs.

What is a sea?

While there is no set definition of a sea, it is generally described as a body of saltwater smaller and shallower than an ocean basin, and may be wholly or partly underlaid by continental crust. In most cases, seas are partially surrounded by land—although the Caspian and Aral seas, for example, are wholly landlocked. The Sargasso Sea is unique in that it is actually an area of the Atlantic Ocean basin bounded by and defined only by ocean currents (see pp.158–159).

There are between 70 and 80 named seas, of which a range of key examples are shown here. Several large areas known as bays or gulfs are seas in fact if not in name—for example, the Gulf of Mexico, which is largely surrounded by Mexico and the US. The largest seas are the Philippine, Arabian, and Coral seas—the latter to the northeast of Australia—while the smallest is the Sea of Marmara, an inland sea off Türkiye that lies between the Aegean and the Black seas.

1. BERING SEA
Connected to the Arctic via the Bering Strait, the northern part of the Bering Sea freezes in winter. It is separated from the Pacific by the Aleutians, an arc of volcanic islands.

2. LABRADOR SEA
An arm of the North Atlantic between Greenland and Canada, this sea is covered in ice in winter and is a major source of North Atlantic Deep Water.

3. SARGASSO SEA
This region of the North Atlantic is surrounded by clockwise-circulating currents that force water upward and toward the center, which is 3 ft (1 m) higher than the outer perimeter.

4. CARIBBEAN SEA
Known for its tropical climate, clear waters, and white beaches, the Caribbean lies between Central and South America, Cuba, Jamaica, and a string of volcanic islands to the east.

5. SCOTIA SEA
A wild and dangerous sea between the Antarctic Peninsula and Cape Horn in the Southern Ocean basin, it funnels the immense Antarctic Circumpolar Current through the Drake Passage.

PTOLEMY'S SEAS

This map shows the world as the Greco-Romans knew it and is based on a description in Alexandrian polymath Ptolemy's book *Geography*, written around 150 CE. It was still used as a reference until the 15th century and features the first use of longitudinal and latitudinal lines. Two principal seas are named, the *Mare Mediteranum* (the Mediterranean) and the *Mare Indicum* (the Arabian), as well as several smaller seas. The ocean in the west remained largely unknown until the "Voyages of Discovery" from the later 1400s.

PTOLEMAIC WORLD MAP

Fierce winds blow from all directions

6. NORWEGIAN SEA
This northerly sea is where warmer, saltier Atlantic waters meet Arctic waters. The warm currents from the south keep it ice-free throughout the year, despite its high latitude.

7. MEDITERRANEAN SEA
Almost completely enclosed between Europe, Africa, and Southwest Asia, the Mediterranean has been the focus of civilization, trade, and warfare for more than three millennia.

8. BLACK SEA
Connected to the Mediterranean, this marginal inland sea has roughly half the salinity of the ocean, averaging only 17 or 18 parts per thousand on its surface.

9. ARABIAN SEA
A large region of the Indian Ocean basin between the Arabian Peninsula and India, this sea was home to some the world's first known sail boats, and lay on a major east-west trade route.

10. PHILIPPINE SEA
The world's largest sea is bounded by its own eponymous tectonic plate and surrounded by deep oceanic trenches, strings of island arcs, and volcanoes.

11. TASMAN SEA
Lying between Australia and New Zealand in the belt of westerly winds known as the "Roaring Forties," the Tasman Sea is known for its storminess.

Evolution and Geology of the Ocean

Earth is the only planet in the solar system with an abundance of water—liquid, ice, and vapor—on and around its surface. Hundreds of millions of years ago, this water covered almost all of the planet, but slowly new landforms emerged and began to move.

> Earth's **ocean and atmosphere** formed around **four billion** years ago

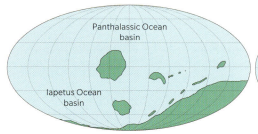

CAMBRIAN (500 MYA)

A world of water
Half a billion years ago, two ocean basins covered Earth's surface. Land area was limited and most of it was situated in the southern hemisphere. None of the modern continents were recognizable.

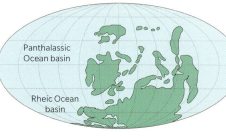

DEVONIAN (400 MYA)

New lands emerge
One hundred million years later, magma escaping from faults in Earth's crust formed new land. Plate tectonics caused these landforms to move around the ocean, which themselves changed size and shape.

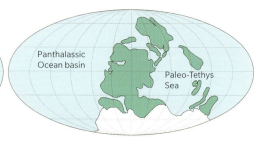

CARBONIFEROUS (300 MYA)

Pole to pole
By the Carboniferous era, Earth's landmasses had converged in a rough line from pole to pole. Ice caps had built up on Gondwana, the southernmost landmass. Pangaea, above it, would eventually drift north and rotate clockwise.

Formation of the ocean

The most likely origin for Earth's oceans is from outgassing, whereby vapor trapped in subterranean rocks and magma was released during countless volcanic eruptions over millions of years, and condensed to form the planet's surface water. It is also believed that the intense bombardment of Earth by ice-laden comets brought more water from space.

Outgassing geyser
Geysers such as this one in Iceland are a smaller-scale version of outgassing. Volcanic activity heats subterranean water, causing it to blast up to the surface as steam and water.

EVOLUTION AND GEOLOGY OF THE OCEAN

Emerging life in the ocean

Life began in the ocean around 4 BYA (see pp.34–37). From single-celled bacteria and archaea, two billion years later the first multicellular life appeared. Landmarks occurred at 575 MYA, during the Ediacaran period, with the arrival of soft-bodied organisms—thought to be early animals—and 540 MYA, with vertebrates, animals with backbones. Larger life forms subsequently emerged, remaining ocean-bound for another 100 million years.

Multiple ammonite fossils form a "death assemblage"

An ocean of ammonites
Prehistoric mollusks with a distinctive spiral shell and bodies similar to squid or octopuses, ammonites lived in the ocean 450–66 MYA.

JURASSIC (150 MYA)

Continents take shape
This, the "Age of the Dinosaurs," saw the recognizable outlines of the modern continents emerge. The Pacific had also assumed most of its present-day form. The other oceans were yet to take shape.

CRETACEOUS (100 MYA)

Shifting into position
Tectonic movement split Gondwana into South America, Africa, India, Australasia, and Antarctica. A similar process to the north formed North America, Europe, and Asia. The Tethys Ocean basin began to disappear.

EOCENE (50 MYA)

A new world
Africa's movement north would see it attach to Europe and create the Mediterranean. India's eventual collision with Asia will form the Himalayas and the Indian Ocean, while the Atlantic Ocean appears in the west.

Why continents move

The Earth's crust—its outer layer—is composed mostly of igneous rock. The crust is fused to the uppermost part of the mantle, a deep layer of mostly semisolid rock, and divided into huge fragments called tectonic plates (see p.18). These plates are constantly shifting around on top of the lower mantle, causing landmasses to move and ocean basins to open and close.

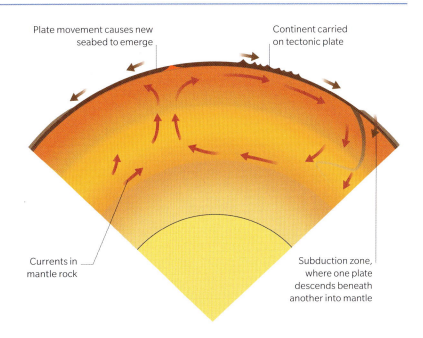

Movement in the Earth
At Earth's core, residual heat from the planet's formation warms the mantle and generates convection currents. Molten mantle rock rises to the surface at volcanoes and spreading ridges on ocean floors.

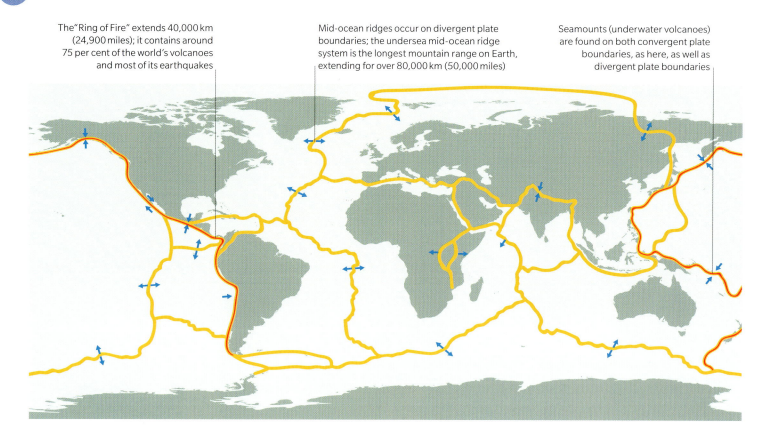

The "Ring of Fire" extends 40,000 km (24,900 miles); it contains around 75 per cent of the world's volcanoes and most of its earthquakes

Mid-ocean ridges occur on divergent plate boundaries; the undersea mid-ocean ridge system is the longest mountain range on Earth, extending for over 80,000 km (50,000 miles)

Seamounts (underwater volcanoes) are found on both convergent plate boundaries, as here, as well as divergent plate boundaries

Tectonic plates

The outer layer of the Earth is around 75–110 miles (120–180 km) thick and is comprised of a cold outer crust fused to the upper mantle. This is known as the lithosphere, which is fractured into a series of very large and some smaller, rigid tectonic plates that are in constant motion relative to one another. Most plates include both ocean and continent; while others are wholly oceanic. Boundaries between plates are irregular and interlocking, and they move at a rate of a couple inches each year.

The size and shape of plates is continually changing as new material is added and old material is consumed. These changes take place at plate boundaries and are the principal cause of earthquakes, volcanoes, hot springs, and heat loss from Earth's interior. At divergent plate boundaries new crust is added (see below), while at convergent margins older crust is consumed or deformed (see opposite). At transform boundaries, the vast plates slide past each other, or sometimes collide to devastating effect, as they did at California's San Andreas Fault close to San Francisco in 1906.

Plate boundaries
The edges of the main tectonic plates are shown here in yellow; arrows indicate their direction of movement at each boundary. The Pacific's "Ring of Fire"—an area of intense volcanic and seismic activity—is shown in red.

Creating the ocean floor

New ocean crust is formed at divergent plate boundaries, where plates move apart from one another and magma extrudes onto the seafloor, more than 1¼ miles (2 km) below the ocean surface. Magma contains iron oxide, which "records" the polarity of Earth's magnetic field. As this reverses every million years or so, the date when each new strip of seafloor was produced can be calculated by whether the direction of its polarity is north or south.

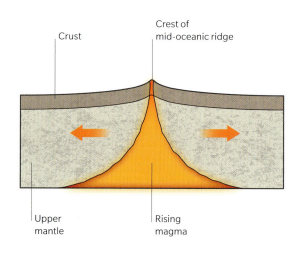

Crust | Crest of mid-oceanic ridge | Upper mantle | Rising magma

New crust and ridges
Molten magma escapes through fractures in the Earth's crust and cools rapidly. Vertical sheets of dark basalt rock (dikes) harden in the fractures and force apart the two sides of newly formed ocean floor. A mid-ocean ridge forms over the fissure.

Subduction zones and crust destruction

Destruction of the ocean crust takes place in subduction zones along convergent plate boundaries, where the great tectonic plates collide. These areas are where older crust is drawn back into the mantle, which happens at the same rate as newer seafloor is produced at mid-ocean ridges. Subduction zones are regions of great heat and pressure, deep-seated earthquakes, and volcanic eruptions.

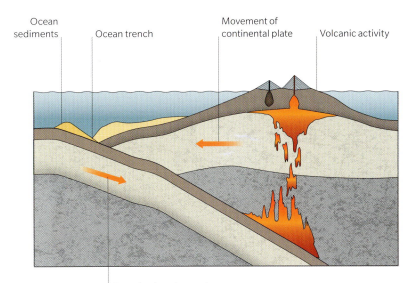

Subduction zone
An ocean plate is forced downward into Earth's interior. The rate of slippage—around 1–3 in (2–8 cm) per year—creates friction, causing rocks to melt and triggering earthquakes.

> " What goes on in those distant depths ... fifteen miles beneath the surface? It's almost beyond conjecture. "
>
> JULES VERNE, *Twenty Thousand Leagues Under the Sea,* 1869

Hell's mouth
Red-hot molten lava at more than 1,800°F (1,000°C) oozes from a mid-ocean spreading ridge to meet ice-cold ocean waters, where it cools into rock formations.

Seawater

Ocean water contains a varied mix of substances, including the one it is best known for—salt. Its chemical composition and temperature both exist in a delicate state of balance.

> **WHAT IS WATER?**
> Water (H_2O) is a compound that exists as a liquid, a solid, and a gas, and is made up of two atoms of hydrogen, and one atom of oxygen. This combination produces a substance with remarkable properties—particularly its ability to dissolve a wide range of substances—that are crucial in determining physical and chemical conditions across the world's ocean and atmosphere.

Composition and exchange

Seawater comprises a mix of nearly 100 different naturally occurring elements, including chlorine, sodium, sulfur, magnesium, calcium, potassium, and carbon—as well as an estimated 20 million tons (18 million metric tons) of gold.

Seawater absorbs gases from the atmosphere and accumulates salts from the land in a continuous process of exchange. Around 3.3 billion tons (3 billion metric tons) of dissolved chemicals are delivered by rivers to the ocean each year, with further deposits from volcanic activity and seafloor vents. The concentration of chemicals in seawater remains relatively consistent, however, because elements are continuously removed from the water by processes including absorption into sediment, and biological processes such as consumption and excretion by marine animals. Photosynthetic marine organisms also output oxygen into the water and atmosphere.

Exchange of elements

The land, ocean, and atmosphere interact together to cycle different gases, salts, and other minerals at different rates through the air, water, and sediment.

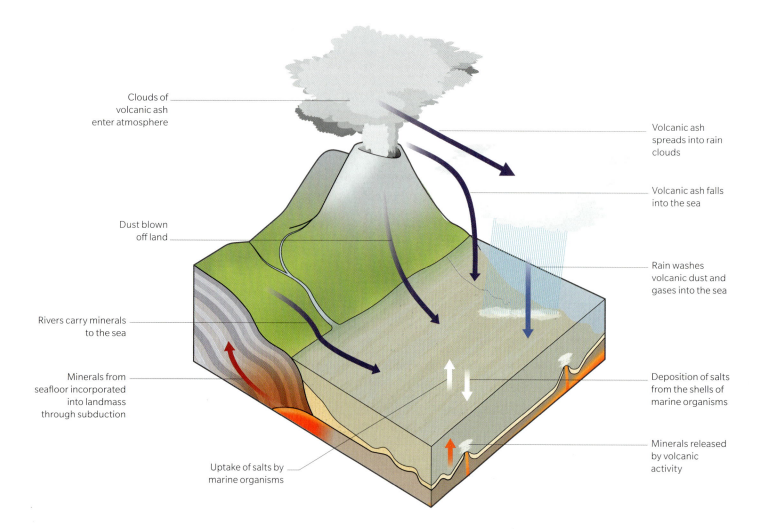

Salt and salinity

Earth's ocean contains more than 5.5 trillion tons (5 trillion metric tons) of different dissolved salts. The total concentration of salts is known as salinity: this averages 3.5 percent for open ocean water (with a range of 3.3 to 3.7 percent). There is enough salt in the oceans to form a layer of salt up to 500 ft (144 m) thick over the entire planet.

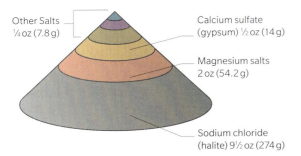

Other Salts ¼ oz (7.8 g)
Calcium sulfate (gypsum) ½ oz (14 g)
Magnesium salts 2 oz (54.2 g)
Sodium chloride (halite) 9½ oz (274 g)

Types of salt
There are different types of salt in seawater, shown here in the salts from 2⅔ gallons (10 liters) of seawater.

Salt pans, Malta
Flat pans are carved from the rock to collect seawater; the water then evaporates, and local people harvest the salt.

Density and temperature

The density of seawater changes depending on its salinity and temperature. An increase in salinity increases the water's density, as does a decrease in temperature.

Changes to salinity occur mainly through the addition or removal of water. Rainfall and input from rivers and from melting sea-ice add fresh water and lower the salinity at local levels. Conversely, as water evaporates or freezes into ice, salinity increases. The surface salinity of the ocean is therefore higher in the subtropics and lower in the Arctic. At great depths, salinity is more or less constant. Denser water will always sink below any lighter layers of water, contributing to the movement of water that feeds deep sea currents (see pp.24–25).

Red Sea water
The Red Sea's water is noted for its high levels of salinity, causing areas of high density. Coral reefs in the Red Sea have adapted to survive in these unusual conditions.

Surface Currents

A network of currents flows long distances in the upper layer of the ocean, transporting vast quantities of water around the globe and performing a vital exchange of heat energy between the tropics and the cooler waters near the poles.

What causes surface currents?

Surface currents are caused by global wind systems, which are fueled by solar energy. Wind blowing across the surface of the ocean causes friction, which moves the water. Due to frictional drag between layers of the water, combined with the Coriolis effect (see right), the current that has been created moves at an angle of up to 45° to the direction of the wind. This, in turn, has the effect of creating large circular systems of currents called gyres (see panel). Surface currents may also be created when water flows from an area of high sea level to an area of lower sea level.

GYRES

There are five ocean gyres, two in the Atlantic, two in the Pacific, and one in the Indian Ocean. These large-scale, circular currents form as a result of prevailing winds and the spin of Earth. The water in gyres piles up in the center, resulting in an area of raised water level. The motion of gyres draws in debris, resulting in "garbage patches."

PLASTIC WASTE CAUGHT IN A GYRE

Principal surface currents
Warm currents (in red) flow along and away from the equator; cold currents (in blue) flow across or around the poles and toward lower latitudes.

The warm Gulf Stream current originates in the Gulf of Mexico and brings warm water to the North Atlantic

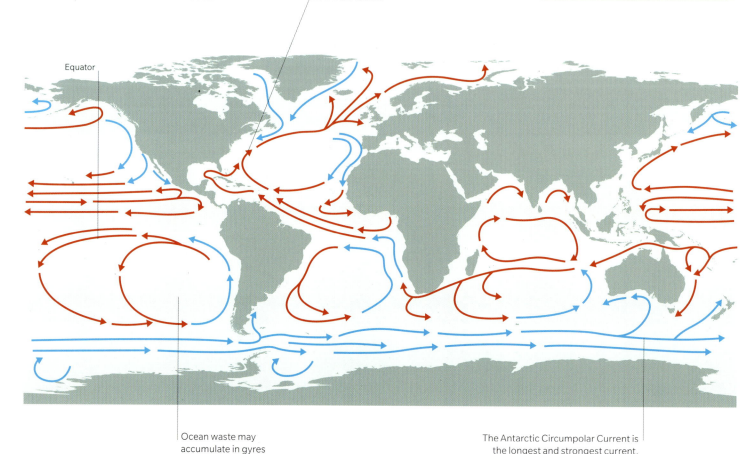

Ocean waste may accumulate in gyres

The Antarctic Circumpolar Current is the longest and strongest current, moving cold water from west to east

SURFACE CURRENTS

Oceanic upwelling and downwelling

As water is moved by surface currents, it may be replaced by water drawn in from below. This upward vertical movement is known as upwelling. It brings nutrient-rich waters to the surface, helping support the growth of plankton. Major upwelling zones exist off the coasts of Peru, Namibia, Mauritania, and western North America, as well as in equatorial and high-latitude belts in the open ocean. Differences in temperature and salinity also cause water to sink or rise, generating currents. This is the major cause of downwelling, where water sinks. These vertical movements link surface currents with deep water currents (see pp.24–25).

Coastal upwelling
Coastal upwelling occurs where longshore surface currents are deflected offshore as a result of the Coriolis effect (see below). This draws colder water upward from below.

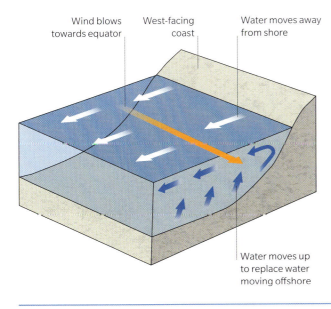

The food chain in action
Upwelling zones where plankton populations are high are busy parts of the ocean ecosystem. Zooplankton feed on phytoplankton (and each other), which in turn attracts schools of smaller fish, and then sharks.

The Coriolis effect

Materials not rigidly attached to the planet's surface—including water, wind, and airborne birds and planes—are subject to the Coriolis effect. This occurs due to Earth's rotation, which exerts a lateral force on unanchored matter. In this way, ocean currents moving to the north or south are deflected sideways from their initial direction of travel—to the right in the northern hemisphere, and to the left in the southern hemisphere.

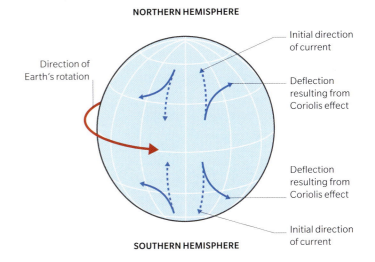

Effect on ocean currents
The Coriolis effect deflects the initial vertical trajectories of ocean currents (dotted lines) into curved trajectories (solid blue lines).

Connected Ocean

Hidden from view, deep, slow-moving, powerful currents connect every part of the planet's ocean, traveling through even the deepest basins and eventually affecting all ocean waters.

It takes **1,000 years** for a **water molecule** to complete one loop of the **global conveyor belt**

Deep currents

Linking with the surface current systems across the world's ocean (see pp.22–23), the deep-water currents below them form a vast network of circulating deep-sea water that transfers energy, nutrients, and sediments around the globe. They are driven by density differences caused by varying salinity and temperatures—cold, deep waters generated in the high North Atlantic and Southern ocean basins flow toward the equator and warm, surface tropical waters flow back. Together, these form the global thermohaline circulation system—or global conveyor belt—that links all of the ocean, from surface to bottom.

Because of this connecting system of currents, it has been argued that, rather than separate bodies, Earth's waters should be thought of as one ocean—a single system where water, energy, and nutrients circulate freely, regulating the climate and supporting marine ecosystems.

Global conveyor belt
Cold, salty water sinks in the North Atlantic. It moves south at depth to the Indian and Pacific oceans, where it mixes with warm water and returns to the surface. Warm surface currents return it to the Atlantic.

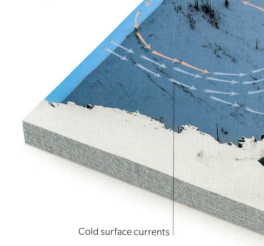

Warm water cools, becomes more dense, and sinks at high latitudes

Warm surface currents

Cold surface currents

Cold Atlantic water merges with cold water formed near Antarctica

Sea ice and deep water

The sea surface freezes solid at high latitudes, especially at the poles during sunless winters. The salt in seawater, however, does not freeze, making the water underneath the frozen layer more saline, and so denser and heavier. This water sinks to the bottom and is carried away around the ocean by deep-sea currents.

Icy seas
These icy seas around Greenland are a source of ultra-cold, dense water. Around 90 percent of the ocean is made up of deep ocean water.

Antarctica and global cooling

The opening of two oceanic gateways (see panel, below), the Drake Passage and the Tasmanian Passage, isolated Antarctica around 34 million years ago by creating the Antarctic Circumpolar Current, which insulated the continent from warmer waters and allowed the first ice sheets to form on it. This changed the pattern of ocean circulation, leading to planetary cooling.

Sea of ice
Each winter, part of the ocean around Antarctica freezes and sea ice extends out 1,900 miles (3,000 km) from the coastline, doubling the continent's size.

Shaping the seafloor

Deep-ocean currents shape the ocean floor via erosion, transportation, and deposition. Continental slopes linking the shallows to the deeper ocean are sculpted by underwater landslides. Sediment-charged currents also deposit material. This can build mounds of sediment over millions of years, and currents can carve terraces and erode deep-sea channels.

Oceanic canyon
Shaped by currents, the Monterey Canyon (center) off the Californian coast cuts over 1 mile (1.6 km) deep into the continental shelf, before sweeping out 300 miles (500 km) to the deep ocean.

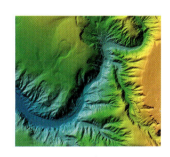

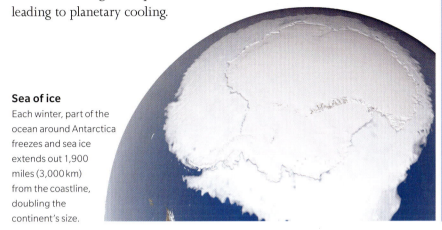

Deep water upwelling to the surface from the deep gets warmed by the Sun's energy at low latitudes

Warm surface currents circulate in north Pacific

Conveyor divides into two branches in the Southern ocean basin

OCEAN GATEWAYS

The ocean is compartmentalized into great abyssal basins and marginal seas, separated by submarine mountain ranges and narrow straits. Both the spill-points across these submarine barriers and the straits between landmasses are known as oceanic gateways. They restrict the flow between basins, so that bottom currents narrow and accelerate, becoming highly erosive, scouring away loose sediment, and even grinding into bare rock. The Strait of Gibraltar between the Atlantic and the Mediterranean experiences some of the swiftest bottom current velocities on the planet.

SATELLITE IMAGE OF THE STRAIT OF GIBRALTAR

Ocean and Climate

The Earth's ocean and atmosphere are closely linked. Together they influence both the daily variations in weather and the long-term changes in the planet's climate. The ocean also acts as a regulator for extremes of temperature.

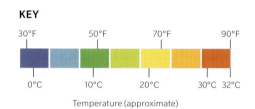

The ocean absorbs **over 30 per cent** of the CO_2 produced by human activity

Temperature distribution

The ocean absorbs and stores heat from the sun. Around ten times more thermal radiation from the sun reaches the equator than the poles, creating a band of warmer waters around the Earth. Surface currents, driven by winds, redistribute this heat from the equator to the poles (see pp.22–23), contributing to the transfer of heat energy to the atmosphere. Without the regulation in temperature that occurs via this transfer, the heat delivered to the tropics would make them uninhabitable and the polar regions would remain permanently covered in ice. Water can absorb or release large amounts of heat energy while changing relatively little in temperature. This helps moderate the climate and is the reason why extremes of heat are greater at the center of landmasses.

Cold to warm
This maps shows the variation in temperature in open ocean waters in March. The Indian Ocean basin (see pp.280–283) has the warmest waters, while the Arctic (see pp.180–181) has the coldest.

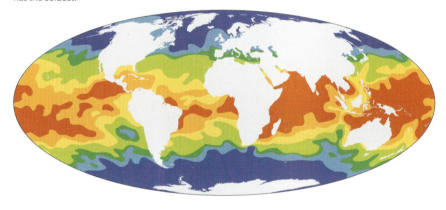

El Niño

El Niño is a disturbance in the ocean-atmosphere system in which ocean surface temperatures become unusually warm. It occurs on an irregular basis every two to seven years and is the single largest contributor to variation in Earth's climate at this timescale, causing changes to global ocean surface temperatures of nearly 1.8°F (1°C) in just one year. During El Niño events, warm waters spread across the equatorial Pacific from Indonesia to South America, where they suppress the upwelling of cold nutrient-rich waters off Peru, altering the ecosystem. While heavy rain and floods batter the Chile-Peru coast, droughts and wildfires rage in Australia, Indonesia, the Sahel, and the US, and the Indian monsoon (see opposite) weakens or even fails. El Niño is the "warm" phase of a wider weather phenomenon, the El Niño-Southern Oscillation (ENSO).

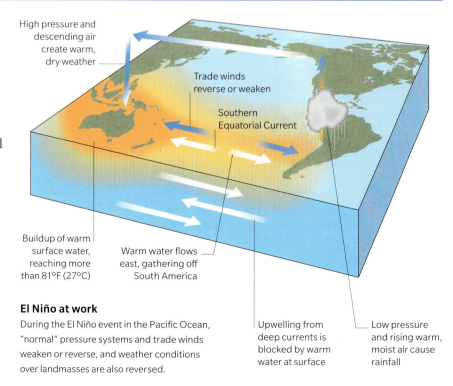

El Niño at work
During the El Niño event in the Pacific Ocean, "normal" pressure systems and trade winds weaken or reverse, and weather conditions over landmasses are also reversed.

OCEAN AND CLIMATE

Ocean–atmosphere exchange

As well as heat and moisture, gases are transferred between the ocean and the atmosphere. The sea surface may both absorb and emit gases; in particular, the ocean holds 50 times more carbon dioxide (CO_2) than the atmosphere, and absorbs more each year. The exchange depends on two key properties. The first of these is the difference in the concentration of gases between air and sea, which controls the direction and rate of gas exchange. The second is the strength of ocean mixing, driven by wind and waves. During storms, more gas is absorbed. Once the gas has dissolved, water turbulence mixes it downward. Other influencing factors include temperature—cold water absorbs more CO_2—and biological activity, as phytoplankton use CO_2 for photosynthesis.

Gathering storm
A supercell thunderstorm gathers over Australia's Queensland coast. As well as increasing absorption of gases, research in 2024 indicates that storms can cause outgassing—the emission of gases such as CO_2 from the ocean.

Monsoon climates

A monsoon is a change in direction in a region's strongest winds, occurring seasonally and causing wet and dry seasons. Monsoon climates affect many tropical regions across both hemispheres, including southern, eastern, and Southeast Asia, northern Australia, southern Africa, and South America. The changes in wind direction are the result of unequal heating of continents and tropical ocean basins. During winter, the continents cool faster than the ocean, as land absorbs and releases heat faster than water. This creates areas of high pressure that force winds to blow outward toward the ocean, leading to dry weather. During summer, these conditions are reversed: the land heats up faster than the ocean, causing air masses to rise, and forming areas of low pressure. These draw in moisture-laden winds from the ocean, resulting in heavy rains and tropical storms.

Storm season
Tropical storms may occur during the summer monsoon season. Generated by ocean warming, they develop when sea surface temperatures exceed 81°F (27°C).

Typhoon Tip, the largest tropical storm on record, measured 1,380 miles (2,220 km) across

Tides

Tides are the regular rise and fall of sea levels, caused by gravitational forces. Water accumulates on the sides of the Earth's surface closest to, and farthest from, the moon, with additional pull from the sun.

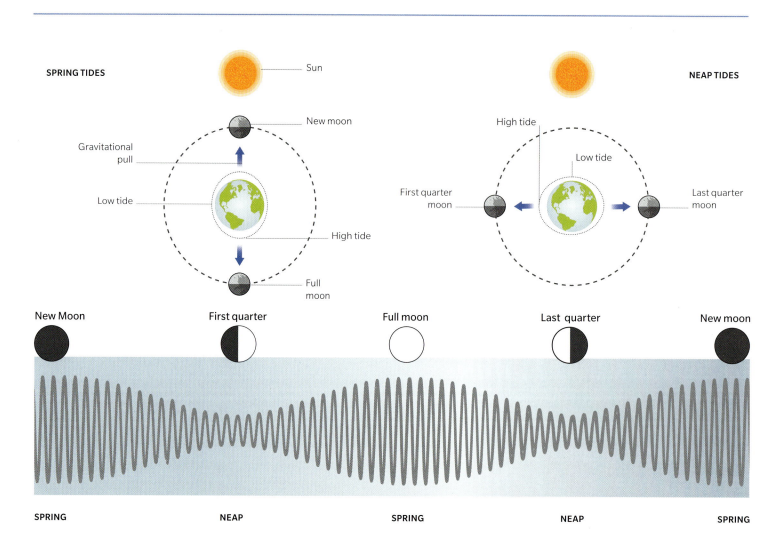

How tides work

The cause of tides was unknown until 1687, when English physicist and mathematician Isaac Newton explained how they are created by the gravitational attraction of the moon and the sun, and by the rotation of the Earth and moon together.

The moon's gravitational attraction is the dominant factor in generating tides, while the sun's gravity is 46 percent as effective. The ocean water is pulled up into a bulge on the side facing the moon, while centrifugal force creates an equal bulge on the opposite side of Earth. When the sun and moon are aligned, the tidal bulge is at its maximum (the spring tide), whereas when they are at right angles, the tidal bulge is at its minimum (the neap tide).

The size and shape of the sea and the position of land influences the timing of high and low tide and the tidal range (the difference in height between high and low tide). This varies from less than 3 ft (1 m) in places such as the Mediterranean to slightly more than 52 ft (16 m) at the Bay of Fundy in Canada.

Tidal patterns

High spring and low neap tides happen just after the sun and moon align. Frictional and topographic restrictions within the sea around which tidal water flows also affect the patterns of the tides.

Wildlife and tides

In the face of rising and falling tides, seaweeds and animals have to adapt to very different conditions. The rising tide floods across sand and mud banks, forcing the wading birds that have just been digging in the mud to move on and allowing fish in to feed on crabs, shrimp, and algae. The seaweeds and animals that stay in place have developed various strategies and adaptations: at low tide, barnacles and limpets withdraw into their protective shells, while seaweeds stay moist thanks to the protective coatings they have evolved.

Tidal power

The energy locked up in tidal movement is estimated at 3,000 gigawatts, however not all of this can be captured. At present, it is believed that 120–400 gigawatts could be generated by tidal barrage power stations. The first such power station, the Rance Tidal Power Station in northern France, opened in 1966. The largest is now the Sihwa Lake Tidal Power Station in South Korea, with 254 megawatts of electricity-generation capacity. Narrow inlets or estuaries with a large tidal range can be good locations for tidal power stations as incoming and outgoing waters are funneled into fast-flowing channels.

Low-tide foraging
A bar tailed godwit (*Limosa lapponica*) forages deep in a mudflat for worms in Norfolk, UK. Many shore birds feed during low tides, which expose large tracts of worm- and shellfish-rich mud and sand flats to search for food.

Harnessing the ocean's power
The tidal power station at Xiushan Island in Hangzhou Bay, China. Upstream, the Qiantang River has the highest tidal bore in the world—a wall of water up to 30 ft (9 m) high that surges upriver with the flood tide.

Tides around the world

Most coastlines experience a semi-diurnal tide pattern—two high and two low tides of roughly the same size per lunar day. Where the two high and the two low tides are markedly different in height, they are known as mixed tides. Some areas, such as the Gulf of Mexico and parts of the Australian coastline, have a diurnal pattern, with only one high and one low tide per day.

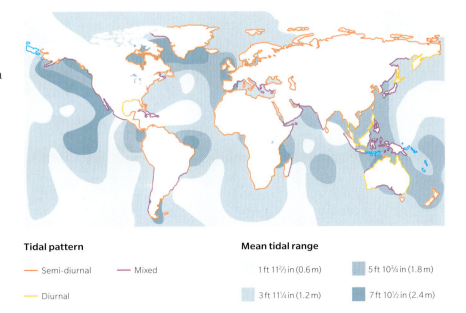

Tidal ranges
The western coast of the Americas experiences both semi-diurnal and mixed tides. The tidal range in the Pacific varies from less than 3 ft (1 m) in places, such as Hawai'i, to around 30 ft (9 m) close to Anchorage, Alaska.

Tidal pattern
— Semi-diurnal
— Mixed
— Diurnal

Mean tidal range
1 ft 11⅔ in (0.6 m)
3 ft 11¼ in (1.2 m)
5 ft 10¾ in (1.8 m)
7 ft 10½ in (2.4 m)

Coastlines

Coastlines, the areas where land meets ocean, are in a constant state of change. The action of waves, wind, and shifting sediment continuously sculpts the shoreline into different kinds of formations, from sweeping dunes to towering cliffs.

Types of coastlines

Coastlines are broadly divided into four categories. Emergent coastlines form due to rising land or falling water level. They can occur when the movement of tectonic plates pushes new landforms up from the seabed (see pp.18–19) and may feature cliffs (see pp.40–41) or sea stacks (see pp.44-45). Submergent coastlines are formed when land drops or sea levels rise. Many of these were formed by rising sea levels around 115,000–11,700 years ago. They include landforms such as partially drowned river valleys (rias), and fjords (see pp.114–115).

Discordant coastlines have bands of different kinds of rock running perpendicular to the water's edge. The varying hardness of the rock causes different rates of erosion, forming bays and headlands (see p.47). Concordant coastlines have different types of rock running in parallel to the water, and may form coves—small, sheltered inlets.

Pacific coast

The Nā Pali coast spans over 15 miles (25 km) of the northern shores of the Hawaiian island, Kaua'i. An emergent coastline, its distinctive cliffs have been formed through erosion by both the Pacific waves and torrential rain waters.

COASTLINES

BEACHES

A beach is a strip of land along a coastline—mainly between high and low tide—made up of loose sediment. This can be sand, gravel, pebbles, or boulders derived from the breakdown of rocks, or broken-up biogenic fragments, such as shell and coral debris. Waves, currents, and storms erode the coastline and grind the fragments into smaller material. This material is then deposited along the coast as beaches, or offshore into deeper water. Winds may also blow sand into dune fields at the back of the beach.

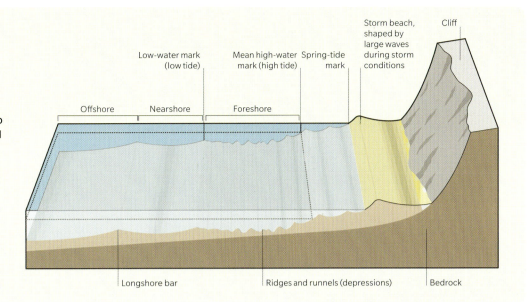

PARTS OF A BEACH

Erosion, transportation, and deposition

Coastlines are shaped by three main interconnected processes: erosion, transportation, and deposition. Erosion is the wearing away of land by the constant movement of seawater and abrasion by the sediment it carries. Transportation is the movement from one area to another of loose material such as sand, silt, and small rocks by moving water. Deposition occurs when waves or tides lose energy and drop this material, which can accumulate to form coastal features such as beaches, deltas, and tombolos (see pp.82–83).

Wearing away, building up
This beach at James Price Point near Broome, in The Kimberley, Western Australia shows both deposition in its beach sands and erosion in its soft cliffs.

Where rivers meet the ocean

Where a river reaches the ocean (or lake, other river, or other body of water), it may slow down, spread out, and lose its ability to carry sediment, causing deposition. Shallow channels, called distributaries, may branch off the main channel, forming a network. Under the right conditions, the deposited sediment gradually builds up to create a distinctive triangular landform extending from the river mouth, known as a delta (see pp.108–111). Parts of deltas may be submerged and others above water, forming islands.

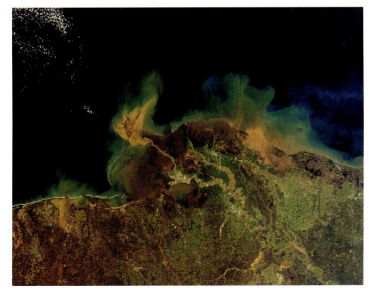

Mississippi Delta
Plumes of sediment-laden water flow out from distributary channels at the Mississippi Delta in the Gulf of Mexico. Here, sediment supply has created a classic "bird's-foot" shape.

Ocean Zones

The ocean is commonly divided into zones that relate to different depths of the water column.

Sunlit Zone
0–650 ft (0–200 m)

Plentiful sunlight and oxygen in the upper ocean support a vast array of plankton, seaweeds, and seagrasses, and the greatest diversity of marine life anywhere (see pp.196–197).

PYROCYSTIS LUNULA ALGAE
Pyrocystis lunula
Dinoflagellates such as these 50–100-µm-long algae make use of sunlight to photosynthesize.

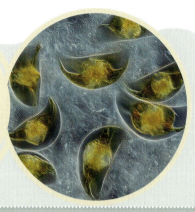

Twilight Zone
650–3,300 ft (200–1,000 m)

Food-producing algae no longer survive at this level (see pp.250–251), but a number of animals have adapted to move, feed, and reproduce in its dim light, many relying on bioluminescence to see.

FIREFLY SQUID
Watasenia scintillans
This tiny predator flashes blue bioluminescent lights in the upper Twilight Zone.

Dark Zone
3,300–13,100 ft (1,000–4,000 m)

Pitch black, uniformly cold, and swept by strong currents, these depths feature sparse fauna—except for the unique oases of life that thrive around hydrothermal vents, cold methane seeps, and deep-sea canyons (see pp.284–285).

BLACK SWALLOWER
Chiasmodon niger
This common tropical deep-sea fish has evolved to swallow prey much bigger than itself.

Abyssal and Hadal Zones
13,100–36,000 ft (4,000–11,000 m)

With abyssal plains, huge sediment fans, and deep trenches on the seafloor, these are the least-known parts of the planet (see pp.298–299). Rare, slow-growing, long-lived animals eke out an existence here.

GRENADIER
Macrourinae
Also known as rattails, these large brown-black fish are some of the few that patrol the abyss.

From the surface to the depths

The ocean zones are characterized by increased water pressure and decreased light as they go deeper, which affects the type and amount of marine life found there. The air above the ocean hosts plentiful marine life including nearly 350 species of sea birds.

ATLANTIC FLYING FISH
Cheilopogon melanurus
This fish can accelerate to 19 mph (30 kph), and glide for 39 ft (12 m) above the water.

MANTA RAY
Mobula
These giant filter feeders are found in warm seas. They mainly subsist on the abundant zooplankton of the sunlit zone.

GREAT BARRACUDA
Sphyraena barracuda
This fast-swimming ocean predator lives up to depths of 328 ft (100 m) in the subtropics.

HATCHETFISH
Sternoptychidae
These small, silvery, bioluminescent fish live in the Twilight and Dark Zones, but come to surface waters at night.

VIPERFISH
Chauliodus
Named for their needlelike teeth and hinged lower jaws, viperfish also migrate to the surface to feed at night.

SPERM WHALE
Physeter macrocephalus
The world's largest toothed predator can plunge to 10,000 ft (3,000 m) to hunt giant squid.

TO THE SURFACE AND BACK

Diel vertical migration (DVM) is a daily, synchronized movement of marine animals between the ocean surface and layers beneath. This is the largest migration on Earth, with trillions of animals ascending to the surface at dusk and returning at dawn. The journey is a trade-off between avoiding predators and gathering food.

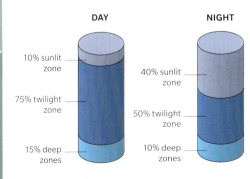

DAY: 10% sunlit zone, 75% twilight zone, 15% deep zones
NIGHT: 40% sunlit zone, 50% twilight zone, 10% deep zones

DAY AND NIGHT POPULATION DISTRIBUTION

> " ... the only other place comparable to these marvelous nether regions, must surely be naked space itself ... "
>
> WILLIAM BEEBE
> *Half Mile Down*, 1934

Ocean Animals

There has been animal life in the ocean for around 560 million years, and it has been both studied and used by humans for millennia for food, tools, jewelery, dye, and more. Today, scientists are still discovering around 2,000 new marine species every year.

White underbelly provides camouflage | **MEGALODON**

c. 3.6 MYA
The largest predatory shark that ever lived—Megalodon (*Carcharocles megalodon*) went extinct due to cooling waters, prey species dying out, and competition from other predators, such as great white sharks.

c. 50–40 MYA
The ancestors of whales lived on land at the beginning of this period as four legged mammals. Within 10 million years they had returned to the ocean and evolved into the whales of today.

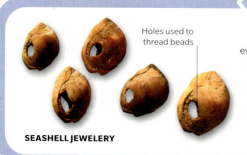

Holes used to thread beads

SEASHELL JEWELERY

c. 73,000 BCE
The earliest evidence of human jewelery is beads made from tusk shells that were discovered in South Africa's Blombos Cave.

c. 1500 BCE
The Phoenicians started making a dye called Tyrian purple from the mucus glands of several species of ocean-dwelling snails.

DYE PRODUCTION

1967
Marine biologist Roger Payne discovered humpback whale song. He went on to release a collection of recordings, *Songs of the Humpback Whale*, which inspired conservation efforts.

WHALE RECORDINGS

1964
The first US deep-sea research submersible, *Alvin*, was commissioned. It made many new discoveries, including life around deep-ocean hydrothermal vents.

ALVIN

2005
Researchers photographed a wild giant squid (*Architeuthis dux*) for the first time. The animal attacked their bait at a depth of 2,950 ft (900 m) off the Ogasawara Islands in the North Pacific.

GIANT SQUID

2016
Deep-sea organisms, such as hermit crabs, squat lobsters, and sea cucumbers were discovered to have ingested microplastics.

OCEAN ANIMALS

> "Men may sail the seas for a lifetime and seldom, if ever, come in contact with the nightmare monsters that inhabit the caves and cliffs of the ocean floor."
>
> WILLIAM OUTERSON, Monster Mix, 1968

Long skull and jaws packed with large teeth

EARLY WHALE

c. 2.6 BYA
Single-celled microbes called cyanobacteria appear in the ocean, producing oxygen through photosynthesis and forming the basis of all life.

CYANOBACTERIA

542–488 MYA
During the Cambrian explosion, many new species of animals appeared in the ocean, including early arthropods called trilobites, echinoderms, and mollusks.

TRILOBITE FOSSIL

FOSSIL SHARK TOOTH

c. 410 MYA
The first fossil evidence of sharklike teeth discovered are from an ancient fish, *Doliodus problematicus*, that lived in the early Devonian period.

c. 530 MYA
The Cambrian explosion led to the evolution of the first fish. They were just 1 in (2.5 cm) long and had no jaws or side fins.

1670s
Dutch microbiologist Antonie van Leeuwenhoek was the first to see microscopic life in water, including bacteria and protozoans, through the use of a rudimentary microscope. He described his findings as "animalcules" (little animals).

ANTONIE PHILIPS VAN LEEUWENHOEK

1831–1836
During his around-the-world voyage on HMS *Beagle*, Charles Darwin studied many marine organisms including corals, fish, barnacles, and plankton.

CHARLES DARWIN'S VOYAGE

JACQUES COUSTEAU

1940–1990s
French oceanographer Jacques Cousteau transformed ocean exploration by inventing the aqualung in 1943, and made marine conservation popular around the world.

COELACANTH

1938
A prehistoric fish called a coelacanth (*Latimeria chalumnae*) was caught after being believed to be extinct for around 65 million years. Another specimen was not seen until 1952.

2019 onward
New hunting behaviors in orcas were first observed, including hunting full-size blue whales, killing sharks to feed on their livers, and preying on leopard seals.

Curved dorsal fin of female

ORCA

2024
Researchers found hundreds of new deep-sea species—including corals, sponges, sea urchins, and crustaceans—during a Schmidt Ocean Institute expedition to seamounts off the coast of Chile.

SEA URCHINS

Plants and Algae

From microscopic phytoplankton to fields of seagrass and forests of kelp, marine plants and plantlike algae provide food and habitats for animal life. They are a vital part of marine ecosystems and are essential to humans' understanding of both the ocean and how life on Earth developed.

SARGASSO SEA

1492
Christopher Columbus was the first person to mention the Sargasso Sea in writing. The sea is named after the Sargassum seaweed that grows there.

ARISTOTLE

c. 350 BCE
Aristotle described bioluminescence, possibly from algae: "When you strike the sea with a rod by night and the water is seen to shine."

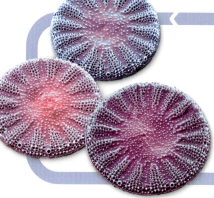

DIATOMS

1703
An "unknown Englishman" observed diatoms—single-celled algae—through a simple microscope. His description was published by the Royal Society of London.

1831–1836
During his voyage on the HMS *Beagle*, Charles Darwin collected various marine plant and seaweed specimens from around the world.

SEAWEED SPECIMEN COLLECTED BY CHARLES DARWIN

SYLVIA EARLE

1956
Sylvia Earle received her master's in marine botany. She would later become one of the world's leading oceanographers, still working to promote conservation in her late 80s.

1890
The Marine Biological Laboratory (MBL) at Woods Hole Massachusetts, which was founded in 1888, started offering general marine botany courses.

SEAGRASS BED

2012
Researchers published a paper explaining the importance of seagrass in storing carbon and protecting the planet from the impacts of climate change.

2013
A sea-star die-off started, causing the population of their normal prey, the purple sea urchin, to boom. This wiped out Californian kelp forests and impacted many other sea creatures there.

GIANT KELP

PLANTS AND ALGAE

> " ... a perfect meadow, a close carpet of seaweed, fucus, and tropical berries, so thick and so compact that the stem of a vessel could hardly tear its way through it."
>
> JULES VERNE, *Twenty Thousand Leagues Under the Sea*, 1870

c. 4.4 BYA
Around 150 million years after Earth was formed, it was covered by an ocean, creating the future habitat of the first life forms.

ANCIENT OCEAN

c. 3.5 BYA
Some of the first life forms on Earth are recorded in stromatolites, deposits of mainly limestone made by cyanobacteria, the first organisms to photosynthesize.

STROMATOLITES

Layers of sediment and minerals

c. 100 MYA
Seagrass—the only flowering plant that can survive permanently submerged in seawater—started evolving. Today, there are around 72 species of seagrass.

SEAGRASS

c. 3 BYA
Fossil records of algae show that they date back three billion years to the Precambrian era.

CYANOTYPE OF ALGAE

1843
Photographs of British Algae by botanical artist Anna Atkins was published, the first book to be photographically illustrated. Atkins made the images—known as cyanotypes—by placing wet algae on light-sensitized paper.

1844
The first record is made of a harmful "red tide" algal bloom in the US, in Florida. This is caused by the alga *Karenia brevis*.

RED TIDE ALGAL BLOOM, US

1887
Marine biologist Victor Hensen first coined the term plankton to describe the tiny organisms—both plant (phytoplankton) and animal (zooplankton)—that drift through the ocean.

PLANKTON

1848
Algologist Isabella Gifford published *The Marine Botanist*. This introduction to British seaweeds and algae contained descriptions of the country's most common species and ways of preserving them.

THE MARINE BOTANIST

2016
Scientists discovered four new species of deep-sea algae, which were collected by a submersible 200–400 ft (61–122 m) deep off Maui, Hawai'i.

DEEP-SEA ALGAE

2020
A study presented a new way of solving the global plastic crisis: creating biodegradable plastics out of algae that are effective but also environmentally friendly.

ALGAL-BASED PLASTICS

A biodegradable water bottle breaking down

CHAPTER 2

Ocean Edges

The places where the ocean meets the land are rich and varied environments that are deeply tied to human history. As tides rise and fall, organisms find ways to subsist in ever-changing conditions, while waves and currents sculpt dramatic landforms.

OCEAN EDGES

Cliffs

Formed where high, rocky ground meets the sea, cliffs provide a habitat for seabirds and other wildlife, and can also contain clues to human history.

Eroded layers
The cliffs of Moher in County Clare, Ireland, reach more than 700 ft (213 m) above the sea. They are formed of sandstone and softer siltstone, which can be seen as clear layers.

The Symplegades
In Greek myth, these deadly cliffs clashed together to kill sailors. Jason and the Argonauts sent a dove ahead of them; after the cliffs crushed the dove and reopened, they rowed through.

The Symplegades cliffs are located in what is now the Bosporus

Cliff dweller
The secretive whiskered auklet (*Aethia pygmaea*) nests in the crevices of cliffs on the Aleutian Islands, Alaska, staying close to the shore.

Facial plumes stand out from the forehead and above the eyes like whiskers

Cliffs are formed by weathering, when rock deteriorates due to natural processes, and by erosion, when the dislodged fragments of rock fall and are carried away. These processes are largely due to the action of the wind and the sea.

The process starts when waves strike the base of the rock. Waves create a hydraulic action: they forcibly compress the air inside cracks and joints, and as the wave retreats the air expands explosively, blowing pieces of rock apart. Waves may dash stones and pieces of rock against the cliff base, in a process called corrasion. They may also abrade the rock by throwing tiny particles of rock against it. These actions produce a hollow at the base of the cliff; as they continue, the rock above the hollow becomes unstable and finally collapses. Over time, the cliff face retreats, leaving a flat stretch of rock called a wave-cut platform at its base.

Many seabirds, including gannets, guillemots, kittiwakes, and puffins (see pp.42–43) nest on rocky cliffs because they can be close to the sea, where they can find food, while being out of reach of most land-based predators, although a formidable aerial predator, the peregrine falcon (*Falco peregrinus*) also nests on cliffs.

On cliffs formed of sedimentary rock, the exposure of deep rock layers can unearth fossils that reveal what lived in the area millions of years ago. For example, in 2022, a fossil enthusiast on the cliffs in Dorset, UK, discovered the 6-ft (2-m) skull of a predatory marine reptile called a pliosaur (*Pliosaurus* sp.), which lived 150 million years ago.

Mythical cliffs
The rocks at the southernmost point of Ireland's Cliffs of Moher are known as Hag's Head, named after the story of a witch who fell to her death while pursuing a warrior who did not return her feelings. According to Icelandic legend, the cliff of Heiðnaberg, or the Heathen Cliff, on Drangey Island was inhabited by a demon. The Icelandic bishop and "folk saint," Guðmundur the Good, is said to have confronted the demon, who pleaded, "even evil must have somewhere to live." The saint allowed the demon to remain.

OCEAN EDGES

Atlantic puffins have been observed carrying as many as 60 small fish at one time

Eyes well adapted to hunting at sea

Prey often consists of small fish

Strong, narrow wing shape is a compromise between the different demands of aerial and underwater flapping

Long-living
Young and nonbreeding Atlantic puffins have comparatively smaller bills and dusky gray—rather than white—faces. Atlantic puffins do not breed until they are at least three years old, but some can live for more than three decades.

Puffins

Fratercula

The three puffin species form a genus within the auk family. They are adapted to deepwater fishing using their wings as flippers, and they are found mainly in the North Atlantic and North Pacific.

Of the world's 24 auk species, the puffins are the most distinctive with their upright, strutting gait; dapper plumage; and large, colorful bills. In tests, the bill of the Atlantic puffin (*Fratercula arctica*) shines brightly when viewed under ultraviolet light. Puffins can see UV light, so the colors serve as a signal to others. All species nest in colonies, often alongside other seabirds. Their relatively small size means they are vulnerable to attack by large gulls and skuas when they come to land, so they nest in crevices or dig burrows in soft cliff-top soil. Most colonies are on islands or steep coastal cliffs, safe from mammalian predators.

In the nonbreeding season, puffins lose their colorful bill-plates, becoming drabber in appearance, and spend their time far out at sea. They catch fish and squid in

The fleshy spike over a horned puffin's eye is ½ in (12 mm) long. It disappears in winter

long feeding dives, "flying" underwater with their short, sturdy wings and reaching depths of 200 ft (60 m) or more. Puffins are also fast fliers, although their proportionately small wings mean flying expends a lot of energy.

Puffins often reuse old nesting sites and reunite with previous breeding partners. These partnerships can span decades, and are reaffirmed each season through rituals of bill rubbing and head bobbing, while exchanging low, growling calls. Pairs share parenting duties and can deliver multiple fish at a time for their chick using a spiny tongue and a specialized bill structure, which allow them to capture a fish without releasing the previous one.

Weather watchers

Sometimes hundreds of puffins in a colony simultaneously take flight with no obvious trigger, circling around the cliff edge in a close-knit flock or heading out to sea together. Known as a "dread," this phenomenon has inspired the belief in coastal communities from Ireland to Alaska that puffins have a special ability to predict changes in sea conditions. In Iceland, the Atlantic puffin was regarded as an expert weather forecaster, while some Indigenous peoples of Alaska believed tufted puffins (*Fratercula cirrhata*) could divert the course of incoming sea storms. A more prosaic explanation of the behavior is that the birds are inspecting their breeding site or practicing their group vigilance, demonstrating to other colony members that they are ready to respond to any threats.

The solemn manner and black-and-white plumage of Atlantic puffins have inspired an association with monks and priests (reflected in the genus name *Fratercula*—meaning "friar" or "little brother"). In Shetland, the Atlantic puffin is known as a "tammie norrie" or "tommy norrie," which, according to *Popular Rhymes of Scotland* (1870), is also a nickname for a young man who is too shy or slow-witted to greet a woman.

Arctic treasures
Northern coastal Norway is known for its midsummer 24-hour daylight along with large colonies of Atlantic puffins, as this 1920s US promotional poster shows.

EXPERT EXCAVATORS
The strong bills and feet of Atlantic puffins are not just for fishing and paddling—the birds also use them to excavate their 3–5-ft- (1–1.5-m-) long nesting burrows. They also reuse old rabbit burrows or even evict rabbits from occupied burrows.

A PUFFIN LOOKS OUT FROM ITS BURROW

HORNED PUFFIN
Fratercula corniculata
This puffin breeds on coasts and islands around the north Pacific. Breeding adults have a very large yellow and orange bill and a black "horn" above each eye.

TUFTED PUFFIN
Fratercula cirrhata
Its mostly black plumage and long yellow tufts of curled feathers make this puffin of the north Pacific very distinctive. It is also the largest puffin species.

44 OCEAN EDGES

Arctic rocks
Many explorers visited Rudolf Island in Russia's Franz Josef Land to survey its harsh polar landscapes, which contain dramatic sea stacks, as depicted in this painting.

Arches and Sea Stacks

As erosion wears away coastal cliffs, it shapes parts of coastlines into bays and headlands. The latter can become arches and sea stacks—striking rock formations carved out by the waves.

> The **Hopewell Rocks** in Canada consist of **more than 20 sea stacks** caused by extreme tides

Arches and sea stacks are rock formations that arise in areas called discordant coastlines, in which bands of harder and softer rock lie at right angles to the coast. The action of waves (see pp.148–151), erodes softer rock such as chalk more easily than harder rock such as limestone. The softer rock is hollowed out to form a crescent-shaped bay in the coastline, while the harder, more resistant rock juts out to form headlands. The headlands may protect the bay from strong winds and waves; the calmer water in the bay allows sand to settle as a beach. Bays have been historically important for humans, because they provided shelter from the sea and could be defended from enemies.

Headlands are more resistant to erosion than bays, but the action of waves can still cause damage over time. Waves approaching the coastline bend to follow the shape of the bays and headlands. The headlands cause the waves to slow, which concentrates their energy. Because the headland is surrounded by the sea on three sides, the waves erode the rock on both of its longer two sides. As waves act on cracks in the rock at the base, the cracks gradually widen and surrounding rock falls away to form a sea cave (see pp.50–51).

Round window
In Iceland's Snæfellsnes Peninsula, the wind and waves have carved away the soft cliff face from the more resistant surrounding basalt rock over centuries to leave this stunning circular window in the Gatklettur arch.

Life on the edge
Birds huddle together on narrow ledges on this sea stack off the island of Bjørnøya, Svalbard, Norway.

The waves enlarge and deepen the cave until it breaks through the other side of the headland, forming an arch. Over time, further erosion wears away the columns holding up the arch until they are too weak to hold up the weight of the roof. The arch may then collapse into the sea, leaving the columns standing as vertical sea stacks. These rock pillars undergo further erosion at their base until they collapse. The short piles of rock that remain are usually only just above sea level at high tide and are known as sea stumps.

Heights and depths

Towering sea stacks, standing well away from the coast, provide seabirds with a safe nesting place and easy access to the water for fishing. The sea stacks at Bandon, Oregon, on the Pacific coast of the US, have been designated as a wildlife refuge for the thousands of seabirds that live on them. They provide an important nesting habitat for birds including the common murre, or guillemot (*Uria aalge*), and the pelagic cormorant (*Urile pelagicus*), which nest on narrow ledges at the sides of the stacks.

The rough, jumbled rocks below the surface around stacks and arches form crevices and pools that can harbor a variety of marine life. New Zealand bull kelp (*Durvillaea willana*) grows thickly around the Nuggets, needle-shaped stacks off the coast of New Zealand's South Island. On the island of Gozo, Malta, the 92-ft- (28-m-) tall Azure Window arch collapsed during a storm in 2017; however, its rocks—now lying up to 130 ft (40 m) underwater—have started to attract algae and aquatic animals. Kicker Rock, in the Galápagos Islands, is a pair of sea stacks with a channel of water between them. The channel is inhabited by animals including Galápagos sharks (*Carcharhinus galapagensis*), Galápagos sea lions (*Zalophus wollebaeki*), and marine iguanas (*Amblyrhynchus cristatus*).

Giants, witches, goddesses, and trolls
According to Icelandic folklore, the Reynisdrangar sea stacks were once trolls that caught passing ships in the night and dragged them to shore. One night, the trolls were said to be still out at dawn; as soon as the sun's rays hit them, they turned to stone. In the neighboring Faroe Islands, a giant and a witch were also said to have been turned to stone in the sunlight, having lost track of time while trying to drag the islands back to Iceland. All that remains of

Summer habitat
In the warmer months, thousands of nesting seabirds inhabit the imposing Icelandic Reynisdrangar basalt sea stacks.

> " Upon Orcade's rocky strand. There stands a man alone. One foot amongst the briny weed. The other one is gone. "
>
> JOHN MALCOLM OF FIRTH, *Orkney*, c. 19TH CENTURY

them, the story goes, are the sea stacks known as Risin og Kellingin, which means "the Giant and the Witch."

Located off the coast of Paphos, Cyprus, Petra tou Romiou is also known as Aphrodite's Rock. In Greek legend, Aphrodite, the goddess of love, was born here. Another legend states that swimming around the rock three times can grant eternal beauty and true love.

Ko Tapu, Thailand, is known for a 66-ft (20-m) limestone sea stack. According to Thai myth, a fisherman brought up a nail instead of fish. When he threw it into the water, it kept returning to him. Eventually, he cut the nail in half and, when he threw it overboard, it turned into the sea stack that still stands today.

HOW HEADLANDS, ARCHES, AND SEA STACKS FORM
Where areas of harder and softer rock meet the sea sideways, wave action erodes the softer rock to create bays. Harder rocks remain as headlands. Waves erode the headland at sea level more slowly until they create sea caves. Over time, the caves deepen and the water breaks through the walls to form arches. When these arches collapse, sea stacks are left behind.

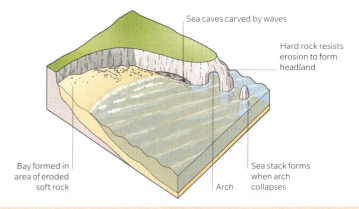

OCEAN EDGES

Distinguished appearance
John and Elizabeth Gould's 1832 painting of great cormorants (*Phalacrocorax carbo*) shows two mature individuals in full breeding plumage, complete with small white feathers on the head and neck that give a frosted appearance.

Facial skin becomes more intensely colored in the breeding season, signaling fitness

Highly streamlined body shape for underwater swimming

Cormorants

Phalacrocoracidae

These slender, prehistoric-looking birds hunt fish underwater, and can be found on coasts worldwide.

There are about 40 species of cormorants and shags. These distinctive seabirds are adapted to swim underwater at high speed, and have webbing between all four toes—this arrangement is known as totipalmate and it provides powerful propulsion. The plumage is permeable to water, so the birds can swim freely at much shallower depths than other diving seabirds, which trap a layer of air against their bodies and so must dive deeper to reach neutral buoyancy.

Hung out to dry
Most species are black or piebald, but many develop striking ornamentations in the breeding season, including crests and brightly colored bare skin, the latter perhaps inspiring the scientific name *Phalacrocorax*—meaning "bald raven." The tufted crest of the Eurasian shag inspired its common name. This species also has a bright yellow lining to its mouth, which breeding pairs display to one another during courtship. Many cormorants

CORMORANTS 49

Pursuit hunter
A double-crested cormorant chases spottail grunt fish. A flock of cormorants will sometimes work together to herd fish into shallow water for easier capture.

DOUBLE-CRESTED CORMORANT
Nannopterum auritus
Widespread throughout North America, this cormorant grows two white fluffy head plumes in the breeding season.

LITTLE CORMORANT
Microcarbo niger
This south Asian species is one of several cormorants that occur inland as well as on coasts.

> " With eager feeding food doth choke the feeder;
> Light vanity, insatiate cormorant,
> Consuming means, soon preys upon itself "
>
> WILLIAM SHAKESPEARE
> *Richard II*, 1595

are highly social and live in colonies. They build large nests from seaweed and guano on rocky crags, and occasionally on the ground or in trees.

Hung out to dry
Lacking the usual seabird waterproofing means that cormorants need to dry out after a dive by standing with wings open. This distinctive pose explains why cormorants have historically been associated with nobility. However, they are also demonized in *Paradise Lost* by John Milton (1667), when Satan takes cormorant form to observe Adam and Eve in Eden. They are known as "greedy"—they sometimes catch fish so large they cannot swallow them—but their fishing skills have seen them adopted as omens of good luck by fishing communities in Norway and other cultures. The accumulated guano of some cormorant species has been used as fertilizer by various cultures for centuries. Inca leaders protected local colonies of guanay cormorants (*Leucocarbo bougainvilliorum*) from disturbance, and this Peruvian species is still the world's most important guano bird in an industry worth more than $1 billion per year. The Galápagos islands (see pp.136–137) are home to the flightless cormorant (*Nannopterum harrisi*), a unique species with undersized wings. With fewer than 2,000 individuals living today, it is one of the world's rarest and most conservation-dependent species.

Elegant icon
This Chinese pendant, thought to depict a cormorant, dates from c. 1050–771 BCE. Cormorants were historically used for fishing in China.

Slender neck is characteristic of cormorant species

Coastal pilgrimage
This woodblock print made c. 1850 by artist Utagawa Hiroshige shows pilgrims visiting the Benzaiten statue and caves on Enoshima Island during Japan's Edo period.

Sea Caves

Found on coastlines worldwide, sea caves are geological formations that have been created over hundreds of thousands of years. They provide homes for animals, shelter for humans, and inspiration for myths and legends.

Also known as littoral caves, sea caves are formed when a weak point in an area of rock is eroded away by waves. Over many years, cracks become larger and pieces of rock fall off until a cave is hollowed out. Sea caves can also be created when rainwater, which contains a weak carbonic acid as a result of dissolved carbon dioxide, dissolves limestone, causing holes to form. Continued erosion can cause the cave to become a sea arch (see pp.44–47) or winding cave system that carves deep into the rock.

Caves can also form through volcanic activity. Anchialine caves, connected to the ocean by underground passageways, can be created when hollow tubes of lava continue to flow beneath lava that has already cooled and hardened. Over many years, caves can reach gargantuan sizes. The Sea Lion Caves in Oregon—one of the world's largest sea caves—have a 1,300-ft (400-m) wave-cut passage with an enormous chamber that measures 310 ft (95 m) long, 160 ft (50 m) wide, and 50 ft (15 m) high. New Zealand is home to the longest sea cave formed by wave action. Already 1 mile (1.54 km) long, Matainaka Cave is still growing.

Regardless of their size, sea caves provide habitats for a range of animals. Their darkness provides cover for fish and invertebrates vulnerable to predation, while delicate creatures such as starfish, corals, and sponges survive better in their calmer waters than in the volatile seas outside. Similarly, they also offer shelter from the elements and from human and animal hunters

The **entrance** to the famous **Blue Grotto on Capri** is just **3 ft (1 m)** high at low tide

to seals, sea lions, and sea birds. Humans have also used caves since prehistory. Archaeologists have found evidence of people living in Torrs Cave, Scotland, from the Iron Age through to the 18th century. An excavation in the 1930s found tools made of bone—probably from the Iron Age—weapons, and pottery from Roman times and the 18th century. Also in Scotland, the Sculptor's Cave on the Moray Firth is believed to be a Bronze-Age burial site.

History and mythology

Pirates and smugglers used sea caves to hide their stolen goods. And, in the 1940s, one inmate from Alcatraz—the maximum-security prison off the coast of San Francisco—hid out in a sea cave for two days after escaping from the jail. Fingal's Cave on the Isle of Staffa, Scotland, is known for its striking, hexagonal basalt columns similar to those of the Giant's Causeway in Northern Ireland. Although they were both created by the same lava flow, legend has is they were once two ends of a bridge with a giant living at each end: Fionn mac Cumhaill in Northern Ireland and his enemy Benandonner in Scotland.

Undersea realm
Saint John's Caves in the Red Sea, Egypt, feature a host of twisting tunnels, caverns, and reefs that scuba divers can explore.

Seal shelter
Human pressures have forced Mediterranean monk seals (*Monachus monachus*) to seek shelter and give birth in sea caves, rather than on beaches.

Fur is typically brown to gray in females

OCEAN EDGES

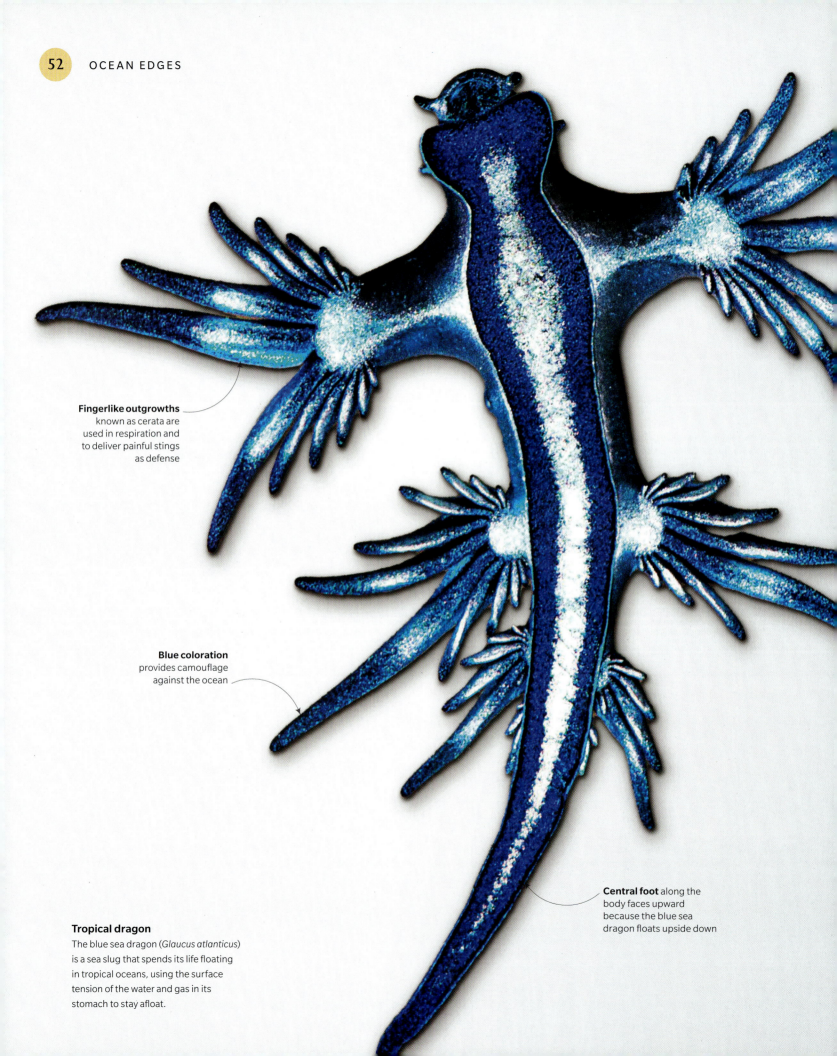

Fingerlike outgrowths known as cerata are used in respiration and to deliver painful stings as defense

Blue coloration provides camouflage against the ocean

Central foot along the body faces upward because the blue sea dragon floats upside down

Tropical dragon
The blue sea dragon (*Glaucus atlanticus*) is a sea slug that spends its life floating in tropical oceans, using the surface tension of the water and gas in its stomach to stay afloat.

Sea shell study
This 1876 watercolour study by British artist Joseph Smith illustrates the range of shapes seen in gastropod shells, from spirals to the egg-shaped cowrie shell (below right).

Gastropods

Gastropoda

Gastropods—snails and slugs—first evolved more than 500 million years ago, and more than 62,000 species live around the world today. Most of these live in marine habitats.

Abalone shell inlays catch the light

Abalone headpiece
The First Nations of Canada's Northwest coast use abalone for sacred objects such as this headpiece worn by leaders of the Tsimshian people.

Gastropods are a class of invertebrates within the phylum Mollusca. All species have similar body parts, including a head with tentacles and a muscular foot positioned just below their gut to help them swim, move across the seabed, or attach themselves to surfaces. Most species breathe using gills—although a few shore-living sea snails have lungs, and some sea slugs have fingerlike structures called cerata along their bodies that enable respiration in a similar way to gills.

Most marine gastropods have a shell, but the various species commonly known as sea slugs have lost theirs in the course of evolution. Unlike bivalves (see pp.262–265), shelled gastropods build a one-piece coiled shell. The shell has three layers: a thin, colored or patterned outer part; a thick, calcium-rich middle layer; and a smooth inner surface, which may be coated with a hard, pearly substance called nacre.

> " Whoever is patient with a cowrie shell will one day have thousands of them."
>
> HAUSA PROVERB, Nigeria

While sea slugs have no shell, many have brightly colored or strongly marked bodies to warn off predators. Some have a foul taste, such as the orange-clubbed sea slug (*Limacia clavigera*). Certain species, such as the blue sea dragon (*Glaucus atlanticus*), even absorb the nematocysts (stinging cells) from prey such as jellyfish or the Portuguese man o' war (*Physalia physalis*), and use the venom from the stings for their own protection.

Distribution and movement

Marine gastropods are found in varied saltwater habitats around the world, from tropical waters to subpolar areas, and from the intertidal zones to ocean depths. In the case of a couple of unnamed snail species found in the Kuril–Kamchatka Trench in the northwest Pacific, these depths can be in excess of 29,500 ft (9,000 m). Some species float freely in the water, feeding on plankton or hunting prey such as jellyfish. Others move around the seabed, using a tongue-like structure called a radula to collect food: for example, abalones (genus *Haliotis*) scrape algae off rocks, and predatory moon snails (family Naticidae) eject toxins through the radula to pierce the shells of other mollusks. Limpets—a varied group of aquatic snails with conical shells—fasten themselves to rock to withstand strong currents, and scrape the surface with their radula to graze on algae. In doing so, large colonies of limpets can wear away sedimentary rock: on the coast of southern England, the

Bright mantle
The colors of the flamingo tongue snail (*Cyphoma gibbosum*) are part of its mantle, tissue that covers the shell in this species and is retracted if the snail is attacked.

The shells take many forms, from flat spirals, as in sundials (family Architectonicidae); to shallow cones, as in limpets (subfamily Patellogastropoda), and elongated spike shapes, as in some conch shells. Some shells have extra growths such as ridges or spikes, as in the tropical sea snails of the *Murex* genus. Many species have a disk of hard tissue called the operculum, which they can pull into the shell opening for protection.

common limpet (*Patella vulgata*) is thought to be responsible for up to 35 percent of the erosion affecting the chalk platforms at the shoreline.

Gastropods and humans

People often use gastropods as food. Limpets, whelks, and periwinkles are pulled from rocks on North Atlantic coastlines. Some species, such as abalone, are fished or farmed in areas such as east Asia, while the queen conch (*Aliger gigas*) is a delicacy in the Caribbean.

Gastropod shells have been used in many cultures as tools or decoration. The Indigenous peoples of the Gulf of Mexico used the curved shells of lightning whelks (*Sinistrofulgur perversum*) as bowls or drinking vessels, and the columella (central column) of the shells to make hammers, chisels, and awls. The Maya of central America, the Indigenous peoples of New Guinea, and Hindu priests in India all used conch shells as trumpets, to summon warriors to battle, or to ward off evil spirits.

More usually, attractive shells or those with a nacreous interior are used for decoration. A dazzling example is the iridescent abalone shell. Among the First Nations peoples of the Canadian Pacific coast, abalone was used as currency and is also a symbol of prestige, added to items such as chiefs' headdresses. First Australians and the Māori see abalone shells as a symbol of protection and healing, while among the Apache of the southwestern US girls wore abalone on their head in the "sunrise" ceremony that marked their passage into womanhood.

In Korea, the art of *najeonchilgi* involves creating complex inlays of abalone and other nacreous shells. In Western countries, abalone has been used to make jewelery and buttons, and as inlays on furniture and on musical instruments such as guitars.

In the eastern Mediterranean, certain sea snails were used to make the dye known as Tyrian purple from about 1570 BCE. The dye comes from mucus secreted by the sea snails to sedate their prey or to defend themselves. The snails would be "milked"—or killed and crushed—to obtain the dye, and it could take up to 12,000 snails to produce just $7/200$ oz (1 g). As a result Tyrian purple was extremely rare and valuable, reserved for use by emperors, nobles, and priests.

FEATHERED CONE SNAIL
Conus pennaceus
Found on the shores of tropical east Africa and the Persian Gulf, this venomous snail can be lethal to humans.

ORANGE-CLUBBED SEA SLUG
Limacia clavigera
This gastropod is found on red algae or kelp along the coasts of western Europe.

GIANT TUN
Tonna galea
This huge, carnivorous sea snail, found on subtropical seabeds, squirts acidic saliva when disturbed.

The **Australian trumpet** gastropod can weigh up to **40 lb (18 kg)**

Spiral shell with distinctive markings

Seafood staple
The mud ivory whelk (*Babylonia lutosa*), a scavenging sea snail found on the shores of the western Pacific, is farmed as seafood in China.

Rocky Shores

Formed by the battering of the ocean waves, rocky shores are home to a variety of wildlife that has adapted to this rugged and hostile environment.

Rocky shores are formed by erosion: over years, strong waves wear away at cliffs, causing cracks. In time, pieces of the cliff break off and the waves distribute the rocky debris along the shoreline. The conditions in this type of coastal habitat can be harsh and unforgiving. Aquatic animals that live on rocky shores must withstand regular exposure to the air and changes in temperature as the tide goes in and out, crashing waves that threaten to wash them away, and high levels of salinity. However, despite these challenges, rocky shores are frequently bursting with life. Seaweeds and lichen are often found here because they tolerate salty conditions, and many species have "holdfasts"—rootlike structures that help them cling on to the rock. These plantlike organisms in turn provide food for grazing species, such as snails and other mollusks, which move across the rocks feeding on them.

Early remnants
It is thought that hunter-gatherers from the late Mesolithic and early Neolithic period frequented rocky sea shores, identified by the piles of seashells (known as "middens") they left behind.

ROCKY SHORES

Rock dwellers
Saltwater mussels (order Mytilida) grip on to rocks using their beards (byssal threads). These act like anchors, preventing them from being washed away by the waves or tide.

Mussels filter more than 7 gallons (25 liters) of water through their body each day

The gaps between rocks also provide places for smaller organisms to hide from predators such as seabirds, which are attracted by the abundance of prey.

Tidal zones

Rocky shores are divided into zones according to the level the tide reaches and the conditions in those areas: the high-tide zone, the spray zone, the middle-tide zone, and the low-tide zone. The spray and high-tide zones are home to animals such as winkles (*Littorina littorea*) and barnacles (Cirripedia) that can survive both in and out of water, as well as plantlike organisms such as algae. Marine animals that need to be submerged longer—such as urchins (Echinoidea) and anemones (Actiniaria)—live in the middle or low-tide zones. The lower zones are also inhabited by larger animals such as crabs, and types of kelp.

Bird life
Spotted shags (*Phalacrocorax punctatus*) nest and breed in large colonies along the rocky coastlines of New Zealand, often lining their nests with seaweed.

Maritime inspiration
Miranda, John William Waterhouse's 1916 painting, was inspired by William Shakespeare's *The Tempest*. Here, Miranda looks on as the ship carrying Ferdinand—with whom she later falls in love—is dashed against the rocks by a storm conjured by her magician father, Prospero. The opening of the play is believed to have been inspired by the 1609 shipwreck of the *Sea Venture* off the coast of Bermuda, en route to the Jamestown colony in North America.

Seals

Phocidae

Seals are a group of aquatic mammals with fur, flippers, and sensitive whiskers. They are superb swimmers and can appear almost human from a distance. Seals evolved from animals related to dogs and bears, and are sometimes nicknamed "sea dogs."

Unlike whales and dolphins, seals can move on land, hauling out onto the shore or sea ice to rest and breed, or to escape predators such as orcas or sharks. They use their front flippers to shuffle along—a process known as "hobbling" or "galumphing." In the water, though, they are incredibly agile. Their back flippers stretch out behind and work like a fish's tail, swishing and rippling to drive the seal forward. All seals are carnivores, feeding on fish, squid, octopuses, crabs, shellfish, or krill, depending on the species and where it lives.

SEALS

Underwater grace
Despite their ungainly movements on land, seals are graceful swimmers. This curious young gray seal twists and turns off the coast of Lundy Island in the UK.

There are 18 living species of "true" or "earless" seals (their relatives, sea lions, belong to a separate group). True seals are found in all the world's oceans, but only a few species live in warmer waters. Most are found closer to the poles, such as the gray (*Halichoerus grypus*) and ringed seal (*Pusa hispida*) in the north, and the Weddell (*Leptonychotes weddellii*) and crabeater seal (*Lobodon carcinophagus*) in the far south. They range in size from the diminutive Baikal seal (*Pusa sibirica*) at 3–4 ft (1–1.3 m) long, to the giant Southern elephant seal (*Mirounga leonina*), which can reach 19 ft (5.8 m) in length and weigh as much as a hippo.

In myth and legend

Perhaps because of their humanlike shape and size, many myths and folktales feature seals that can become humans, and vice versa. The Greek sea nymph Psamathe transformed herself into a seal, while the Scottish selkie could remove its seal skin to walk among humans. In Inuit legends, the fingers of the sea goddess Sedna were cut off and became seals and walruses.

Soapstone is often used by Inuit sculptors

Front flipper has five claws, a trait unique to true seals

Seal sculpture
This soapstone carving of a child and seal is thought to be by Inuit artist Peter Pitseolak (1902–1973), who documented the Inuit way of life through photography, art, and sound recordings.

> " Then the air lifted me up alive ... and carried me far over the seal's bath. "
>
> ANGLO-SAXON RIDDLE, "Barnacle Goose", 11TH CENTURY

SENSITIVE WHISKERS

All seals have whiskers, which they use to sense prey moving through the water, allowing them to hunt in the dark. Ripples in the water make the whiskers move, and nerves around their roots pick up the signals to tell the seal which way prey is heading.

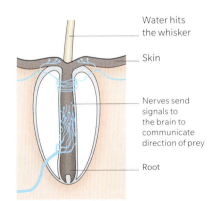

Water hits the whisker

Skin

Nerves send signals to the brain to communicate direction of prey

Root

CROSS SECTION OF A SEAL WHISKER

RIBBON SEAL
Histriophoca fasciata
Immediately identifiable by its wide, pale-banded markings, this seal lives in the Arctic and the northern Pacific oceans.

HAWAIIAN MONK SEAL
Neomonachus schauinslandi
Native to the Hawaiian islands, this endangered seal has a population of fewer than 2,000.

OCEAN EDGES

Ancient species
Ammonites were prehistoric marine mollusks, related to present-day cephalopods such as cuttlefish, octopuses, and squid. They were born with small shells and "built" new chambers as they grew, moving their bodies into the new area and sealing off the old one.

Walls known as septa may have strengthened the shell so it could withstand the pressure of deep water

Body chamber at the end of the spiral is where the ammonite lived

Chambers could be filled with gas or water to sink or float

Coastal Fossils

Shorelines such as Mons Klint in Denmark, Calvert Cliffs in Maryland, and the Jurassic Coast in Dorset, UK, hold numerous fossils—the preserved remains of marine plants and animals that died millions of years ago. These range from coiled ammonites and squid-like belemnites to the skeletons of huge prehistoric sea creatures.

Although fossils of some terrestrial organisms can be found on shorelines, most were formed when areas that are now land were under the sea. Over millions of years, layers of sand, mud, and the crushed shells and skeletons of sea creatures settled on the seabed. As further deposits were laid down, the sediment was compressed, eventually becoming sedimentary rock such as sandstone, limestone, and shale. Dead animals that sank to the seabed and were quickly covered by sediment are most likely to have been fossilized. After their softer body parts rotted or were eaten, harder body parts such as shells, bones, and beaks were buried between the layers of mud and debris.

The most common form of fossilization is permineralization. Water containing dissolved minerals such as calcium carbonate or silica soaks into the remains, gradually replacing the organic material with these minerals. Other fossils—often plants, such as *Coniopteris*, a fern from the Jurassic period—are preserved by being crushed between deep layers of sediment. Fossils may also form when soft-bodied animals such as mollusks dissolve but leave a space in the rock, and this space is later filled with sediment, producing a "cast" of the animal's shape.

A More Ancient Dorset
This 1830 watercolor by geologist Henry De la Beche is based on evidence from fossils found at Lyme Regis, Dorset, on the UK's Jurassic Coast.

" ... it is to fossils alone that the birth of the theory of the earth is due ... "

GEORGES CUVIER,
Discours sur les Révolutions du Globe, 1822

Fossils are often exposed when weathering or land disturbances cause pieces to break off from rock faces, revealing the interior of sedimentary layers. On coastlines, this exposure may result from damage to cliff faces by wave action, rises or falls in sea level, large-scale changes such as the movement of tectonic plates or the retreat of ice sheets, or human activity such as quarrying.

Revealing the past
Measuring certain radioactive elements of the rock in which a fossil is buried can help indicate its age. Conversely, some fossils can be used to date the surrounding rock based on their known age. Ammonites are particularly useful for this purpose because they evolved rapidly during their 140 million years of life, and their presence can be used as a guide to date both rock and other nearby fossils to within about 200,000 years of geological time. Fossils can also help reveal the progress of continental drift.

For example, marine fossils dating from around 470 million years ago have been found at the summit of Mount Everest, showing that the area was once seafloor, before the Indian subcontinent collided with Asia to form the Himalayas.

Jurassic fossils, 1908
This illustration of fossils is from a German encyclopedia, and includes an ammonite (top right) and a bryozoan, a small invertebrate (center bottom).

Curved and ridged oyster fossils (Gryphaea), these are commonly known as "devil's toenails"

Night environment
At night, rock pools can appear quiet and serene. However, many of the animals that live in this habitat are active at night, and below the surface species such as crabs, starfish, and snails may be moving around and feeding.

Rockpooling became popular in the 19th century when railroads enabled easy travel to the coast

Exploring tide pools
American artist William James Glackens' 1906 oil painting *Beach at Dieppe* shows people enjoying a day searching rock pools on the French coast.

Rock Pools

Also known as tide pools, rock pools are coastal habitats that form as ocean tides go in and out. This ever-changing environment is harsh, but many species have adapted to thrive here.

Rock pools are a common marine habitat found on coastlines around the world. However, it is no easy feat for living things to survive in them. At high tide, the rocks are submerged and cool seawater floods in, bringing oxygen and nutrients. At low tide, the water drains away and animals must compete for space in small pools. On hot days, the water evaporates, making the pools even saltier, and temperatures rise. The pounding of waves can also wash animals out to sea.

To survive in this habitat, animals have developed various adaptations. When the tide retreats, the common blenny (*Lipophrys pholis*) may sometimes find itself out of water. This remarkable fish can survive for hours among moist seaweed or wet crevices in the rock, and is able to breathe air until the tide comes in again. Many species of crab have specialized gills that allow them to adapt to changing levels of water salinity. Common limpets (*Patella vulgata*) clamp themselves onto rocks so the crashing waves do not wash them away. At high tide, they release their grip and move around, grazing on algae.

Rockpooling
Rockpooling is a traditional seaside pursuit—looking for interesting marine life in pools exposed at low tide. Starfish, anemones, fish, crabs, snails, and sometimes even the egg cases of sharks and rays can be found. Around the time of full or new moons, the spring tide—which is stronger and goes out farther—reveals lower habitats and a wider variety of species. Many species synchronize with lunar activity. Some sea urchins are more likely to spawn during a full moon, while oysters open their shells more widely at a new moon.

Empty shells are used by hermit crabs

Rock pool dwellers
The common hermit crab (*Pagurus bernhardus*) is often found in rock pools, but it can survive in waters as deep as 500 ft (150 m).

OCEAN EDGES

Sea Urchins
Echinoidea

The sea urchin is a type of echinoderm—a group of animals that includes sea cucumbers and starfish. The name comes from the Greek words *ekhinos* (hedgehog) and *derma* (skin) because the urchin's spiny body resembles a hedgehog.

> The **red sea urchin** can live for more than **100 years**

There are more than 900 species of sea urchin. These spherical creatures are covered in spines that jut out in all directions. Urchins have a complex nervous system—but no brain—and their bodies exhibit radial symmetry, consisting of five symmetrical parts surrounding a central point. Urchins use moveable spines and appendages called tube feet to crawl around and find food. The tube feet have light-sensing cells, and the sea urchin is one of the only animals that has no eyes but can still see. Urchins inhabit coral reefs, intertidal zones, kelp forests, and seagrass meadows. They feed on seagrass, kelp, and algae using a sharp five-piece mouthpart known as Aristotle's lantern. Early zoologists used this term based on Aristotle's studies of urchins in the 4th century BCE.

Regeneration
If a sea urchin loses a spine, it has the remarkable ability to regenerate the missing part. It can also regenerate its tube feet and other organs.

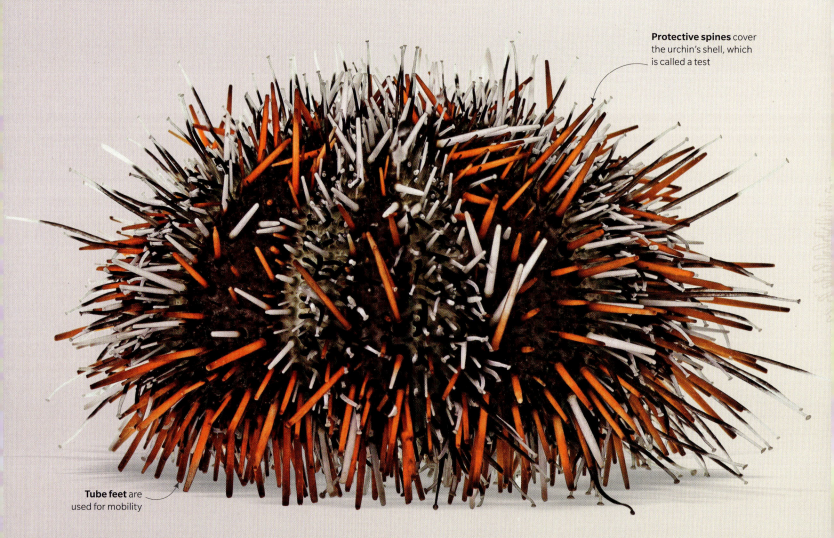

Protective spines cover the urchin's shell, which is called a test

Tube feet are used for mobility

SEA URCHINS

Too much algae can unbalance ecosystems, and the urchins' algae diet can prevent reefs being smothered. Sea urchins are also preyed on by fish, starfish, and sea otters.

In folklore and history

For centuries, people placed fossilized sea urchins on windowsills before a storm. They believed the fossils were "thunderstones" created by lightning. Since lightning does not strike the same place twice, thunderstones were said to give protection from the storm. Another piece of folklore involves the sand dollar, a close relative of sea urchins. If an intact sand dollar is broken open to reveal the mouthparts, they look like five tiny "doves of peace" that were supposed to bring peace into the world.

Some of the hundreds of species of sea urchin—but not all—are edible. During the 20th century in Japan, sea urchins became more popular as a sushi ingredient, and the practice of eating them gradually spread across the Pacific and the Atlantic. The urchin's orange gonads, known as *Uni*, are considered a delicacy in Japan.

In northern California, the native Pacific purple sea urchin (*Strongylocentrotus purpuratus*) population exploded after a reduction in numbers of the starfish that usually eat them. Over five years, the purple sea urchin population increased by as much as 10,000 percent in some areas. These voracious feeders consumed kelp forests at an alarming rate, causing devastating kelp loss. The red abalone, as well as many other marine animals, are struggling to survive without the kelp for food and shelter.

Sea urchins and sand dollars
Sand dollars, various burrowing species of urchin, are flat with short spines, unlike long-spined urchins, as seen in this lithograph from Ernst Haeckel's 1904 book *Kunstformen der Natur*.

EDIBLE SEA URCHIN
Echinus esculentus
Growing up to 6 in (15 cm) in diameter, this urchin inhabits coastal areas of western Europe. It feeds on seaweed and barnacles.

PURPLE SEA URCHIN
Paracentrotus lividus
This urchin gnaws out protective hollows in rocks with its spines and teeth. The specimen shown here has a limpet shell on it.

BLACK SEA URCHIN
Arbacia lixula
Black sea urchins are commonly found in the Mediterranean Sea and feed on coralline algae.

> Some species of urchin have been observed adorning themselves with seashells and wearing them like a hat

CLEANING AND STINGING

Sea urchins and starfish have structures called *pedicellariae* which are small, flexible stalks with three pincerlike jaws (valves) at the end. The urchin uses these moveable structures to clean itself and remove debris, hunt tiny prey that drifts past in the water, and nip any predators in self-defense. Some *pedicellariae* contain venom glands and can sting.

SPINES AND *PEDICELLARIAE*

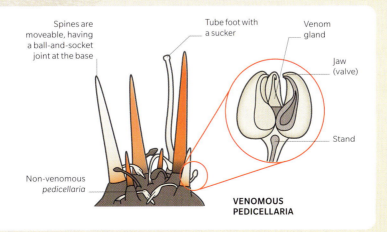

CRAB ANATOMY

Most crabs have a hard carapace and large claws or pincers to protect them from predators. All crabs have five pairs of legs (including their front claws) to help them move around; the hermit crab's last two pairs of legs are much smaller because they are used to hold on to its shell, a salvaged foreign object. The last segment of a hermit crab's abdomen is called a telson, which has appendages called uropods on either side. Together, these help it grip tightly to its borrowed shell.

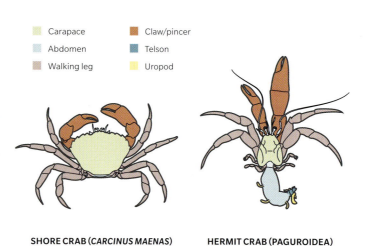

SHORE CRAB (*CARCINUS MAENAS*)   HERMIT CRAB (*PAGUROIDEA*)

Common lobster
Known as the common lobster or European lobster, this species (*Homarus gammarus*) is found in the eastern Atlantic, Mediterranean, and Black Sea. It has large claws and can grow to around 3 ft (1 m).

> " No mollusk or crustacean can produce any natural voice or sound."
>
> ARISTOTLE, *History of Animals*, c. 350 BCE

Crustaceans

Crustacea

From large crabs and lobsters to the tiny organisms that are part of the zooplankton, crustaceans are found throughout the ocean. These distinctive animals feature heavily in myths and stories in cultures around the world.

Crustaceans are a type of arthropod with a hard exoskeleton made of chitin, a segmented body, at least four pairs of jointed limbs, and two pairs of antennae. There are more than 65,000 aquatic species of crustacean. Ninety percent of these live in marine habitats, while the remaining 10 percent are found in fresh water. Woodlice (sow bugs and pill bugs, suborder Oniscidea) are the only fully terrestrial species.

There are various notable types of crustaceans. Decapods, such as crabs, lobsters, crayfish, and shrimp, have ten legs and a more complex nervous system than other crustaceans. Stomatopods, or mantis shrimp, have a shell that does not cover the back of their thorax, gills, or the sides of their abdomen, and are known for their specialized claws that spear or punch prey. Isopods have a flat body and

seven pairs of legs, while amphipods, such as sand hoppers and sand fleas, look like tiny shrimp. Euphausiids, also known as krill, are plankton. Crustacea also includes barnacles, which are sessile, meaning they are fixed to one place with no means of self-locomotion.

Little and large

Crustaceans vary widely in size. The largest species is the Japanese spider crab (*Macrocheira kaempferi*), whose leg span can reach around 12½ ft (3.8 m), with a body of around 16 in (40 cm). This species can weigh 42 lb (19 kg). The American lobster (*Homarus americanus*), meanwhile, is not only the heaviest crustacean in the world but also the heaviest arthropod, at around 44 lb (20 kg) in weight. At the other end of the scale, the tiny *Stygotantulus stocki* species, thought to be the smallest crustacean, measures around $\frac{1}{254}$ in (0.1 mm).

Some of the smallest crustaceans are vital to their underwater environment. Krill play a huge role in the ecosystem by providing a food source for larger species of marine animals, including whales.

Zodiac symbol

This Romanesque depiction of the Cancer sign of the zodiac—the crab—in *The Hunterian Psalter* illuminated manuscript was created around 1170.

Six legs are depicted—crabs usually have 10

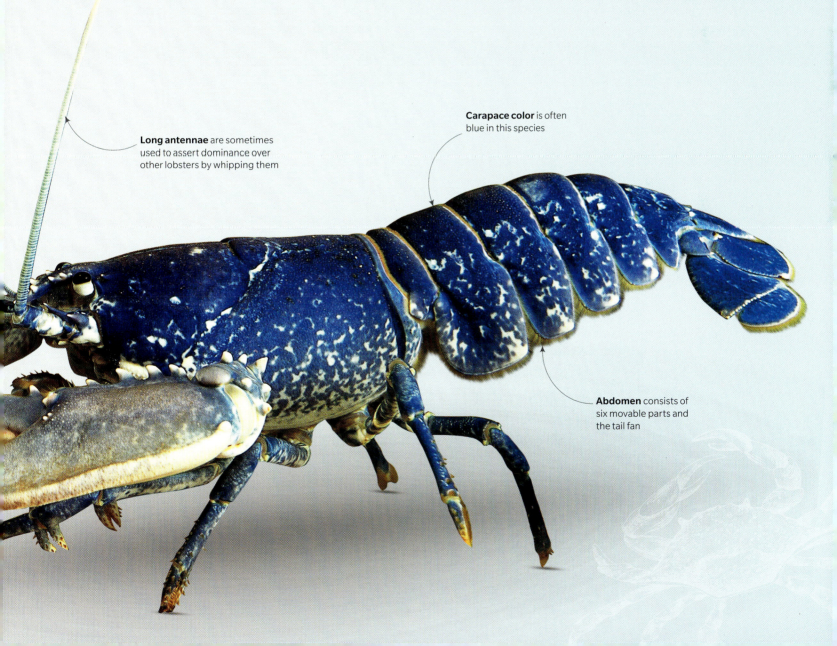

Long antennae are sometimes used to assert dominance over other lobsters by whipping them

Carapace color is often blue in this species

Abdomen consists of six movable parts and the tail fan

Winged trickster
In the Indian tale of "The Heron and the Crab," a heron tricks a shoal of fish into letting it carry them to another pond where it eats them. The crab cuts the heron's head off to save itself from being eaten, as shown in this 8th-century Arabic depiction.

Defense and protection

A crustacean's tough outer shell protects its soft body from predators, but this armor does not grow larger to accommodate the animal's growth. Instead, the animal molts this outer protection and grows a new one. The molting period is dangerous because the animal is left without protection until the new exoskeleton hardens. For coconut crabs (*Birgus latro*), this can take around one month. Lobsters (Nephropidae), among other crustaceans, can regenerate their body parts. This means that, when under attack, they can amputate their own limb or claw to allow them to get away, and then regrow it later. Unlike other crab species, hermit crabs (Paguroidea) do not grow their own shells. Instead, they find and use shells left behind by dead mollusks. If a hermit crab needs a larger shell but cannot find the right size, it waits nearby until others arrive. Eventually, there are enough crabs to switch shells and each leaves with one of the right size. Pom pom crabs (*Lybia tessellata*), have developed an unusual defense and hunting method. They hold tiny anemones in their claws like pom poms, and use them to catch small particles drifting through the water, which they can then eat. The anemones also sting, meaning the crabs can use them for self-defense.

Gods, ghosts, and monsters

In Greek myth, Carcinos was a giant crab that was sent by the goddess Hera to distract Heracles (called Hercules in Roman stories) while he was

Star-shaped holes punched in the card so light could shine through

Cancer, the crab
This 1824 depiction of the constellation of Cancer was part of *Urania's Mirror; or, a View of the Heavens*, a set of 32 astronomical star chart cards.

fighting the Hydra, a giant monster with multiple snake heads that could kill with its poisonous breath. The crab managed to nip at Heracles' feet but could not do much damage before the hero stood on the crab and crushed it to death. Hera memorialized Carcinos by putting it in the sky and turning it into the constellation called Cancer.

In Hawaiian folklore, a hungry god tried to grab an alakuma crab (*Carpilius maculatus*) to eat. The crab did not want to become a meal so it pinched the god back in self-defense, drawing blood with its sharp pincers. Undeterred, the god kept trying to catch the crab, leaving bloody fingerprints on its shell. It is said that this is how the alakuma crab got its spots.

The shells of heikegani (*Heikeopsis japonica*) found in the waters around Japan have the unmistakable image of a human face on their back. Legend has it that these crabs are the reincarnated souls of samurai warriors that drowned in a battle at sea, giving them the nickname "samurai ghost crabs."

Technology and food

Crustaceans have been the inspiration behind engineering breakthroughs. Scientists have modeled swimming robots on krill, and self-digging robots on burrowing mole crabs. Crustaceans are also helping to solve biofouling: chitosan, a product made from crustaceans' chitin exoskeletons, stops the buildup of microbes that attract barnacles and has been used in anti-fouling coatings on ship hulls.

Many crustaceans are eaten by humans, particularly crabs, lobsters, and shrimp, with millions of tons harvested each year. In 19th-century North America, lobster was cheap and abundant, mainly eaten by workers, servants, and prisoners; it was also used as fertilizer and fishing bait. In the late 1800s, the arrival of canning and railroads caused an increase in demand, pushing up prices and making it a luxury commodity.

Perfect partnership
Various species of cleaner shrimp have a mutually beneficial partnership with some fish. The fish is kept free of parasites while the shrimp feeds.

" The hermit-crab grows spontaneously out of soil and slime, and finds its way into untenanted shells. "

ARISTOTLE, *History of Animals*, c. 350 BCE

AMERICAN LOBSTER
Homarus americanus
This is the world's heaviest crustacean. The largest recorded individual was over 3 ft (1 m) long and weighed more than 44 lb (20 kg).

MANTIS SHRIMP
Stomatopoda
Some mantis shrimp have a punch strong enough to break glass, which they use to overcome hard-shelled prey.

ANTARCTIC KRILL
Euphausia superba
There are around 700 trillion krill—weighing over 450 million tons (400 million metric tons)—in the Antarctic Ocean, which are a food source for many other species.

Rocks and pebbles
The pebble beach at Ribeira da Janela, Madeira Island, Portugal, is formed from abraded fragments of the volcanic rock that are typical of this region.

Smooth, rounded edges are caused by abrasion by wave action

Rocky Beaches

Beaches formed from pebbles are common in Europe, but are also found around the world, from the coasts of the US to Birdlings Flat in New Zealand.

Like sandy beaches, rocky beaches are created by "constructive" waves, which deposit more material at the shoreline than they take away. Waves that can carry rocks and pebbles, however, are higher-energy than those that deposit sand. The waves expend their force in throwing the stones on to the shore, but the stones absorb the energy of the backwash so it is not strong enough to pull the stones with it; as a result, rocky beaches are narrower and steeper than sandy ones.

between the rocks, these beaches give shelter to plants such as sea kale (*Crambe maritima*), and invertebrates such as the scaly cricket (*Pseudomogoplistes vicentae*). Birds such as the ringed plover (*Charadrius hiaticula*) nest on the beach, their eggs camouflaged among the rocks.

> " The beach was a desert of heaps of sea and stones tumbling wildly about, and the sea did what it liked ... "
>
> CHARLES DICKENS, *A Tale of Two Cities*, 1859

Harsh environment
The stones come from weathered material shed from cliffs and carried along the coast by a wave process called longshore drift. Attrition by the sea wears away the pieces of debris, causing them to become rounded. During storms, a "storm beach" may be formed as powerful waves cast large stones and rocks up to the high-water mark of the shore. Although a hard, dry habitat, because there is no soil and water drains away

Size gradation
The stones on rocky beaches vary in size. Strong waves throw the largest rocks to the back of the beach; the smaller, lighter, most abraded pebbles land close to the shoreline.

Under the rocks
Ruddy turnstones (*Arenaria interpres*) flip the stones on shingle beaches to find prey underneath. These surprisingly strong birds can lift rocks as heavy as they are.

OCEAN EDGES

Botanical study
This partly unfinished drawing and watercolor of a yellow horned poppy, was created by Austrian painter Moritz Michael Daffinger c. 1840.

> "Her leaves are glaucous green and hoar,
> Her petals yellow, delicate."
>
> ROBERT BRIDGES,
> *The Sea-Poppy, c.* 1904

Yellow Horned Poppy

Glaucium flavum

This bright yellow perennial is found on the coasts of the UK, Europe, and north Africa. It is perfectly adapted to harsh, dry conditions such as shingle.

Bowed head suggests sleep

These coastal poppies, which often grow in small clumps, are usually found on coastal sand dunes, shingle, or cliffs, and are also known as sea poppies. They are adapted to tolerate both dry and salty conditions and are never found inland. They can grow to around 3 ft (1 m) tall and have waxy leaves, which help them retain water, and are covered in short hairs to protect them from salt. They also have long roots.

Burst of color
For most of the year, these plants are identified by their hairy, blue-gray leaves. In summer, they explode with color when their bright yellow flowers bloom, attracting bees, butterflies, and other pollinators. Although the plants bloom throughout the summer, each flower only lasts for around one day. Once they have finished flowering, they develop hornlike seed pods, which can grow up to 12 in (30 cm) long. When they burst, hundreds of tiny seeds are scattered across the landscape.

Like other species of poppy, *Glaucium flavum* is toxic when ingested. Despite this, it has been used in traditional medicines for hundreds of years to treat pain, coughs, and respiratory illnesses. Poppy seeds from both this and the related opium poppy were used as a cure for insomnia, with the latter becoming associated with Morpheus, the Greek god of dreams.

Blue-grey leaves are covered in tiny hairs

Well-traveled species
Yellow horned poppies are an introduced species in the US, where they have been growing since the 17th century. Thomas Jefferson planted some in his garden in 1807.

The dreamer
This sculpture by Fyodor Tolstoy (1783–1873) is of the Greek god Morpheus. His association with poppies led to the naming of the drug morphine, which is derived from opium poppies.

Tiny thimbles
Thimble jellyfish (*Linuche unguiculata*) live in lagoons in tropical and subtropical waters. Here, they are grouped together over a reef in the Philippines.

The sheltered conditions of the Venice Lagoon, northern Italy, gave rise to the powerful Venetian Republic from the 7th century CE

Coastal Lagoons

A coastal lagoon is a shallow body of water that is partially separated from the ocean by a land barrier. These sheltered areas can provide safe habitats for various species of marine animals, particularly juveniles.

Coastal lagoons make up more than one tenth of the world's coastlines. These formations develop along flat, gently sloping coastlines, separated from the ocean by a barrier, such as a coral reef, sandbar, spit, or island (see opposite).

Conditions in lagoons can vary widely. The water in shallow lagoons largely cut off from the sea may be very saline (salty), while lagoons that are partially fed by rivers and runoff from the land may have brackish water (a mixture of seawater and freshwater), and those with several inlets between barrier islands may be regularly replenished with seawater. Shallow lagoons may form swampy wetlands, while deeper ones may be more like lakes or bays. Coastal lagoons are constantly changing, and over time some may disappear—either filled in by sediment, or turned into a bay as their land barrier is eroded.

Protected environment

The land barriers separating lagoons from the open ocean create a protective habitat for wildlife. As a result, lagoons host a wide range of species, some of which are specially adapted to the conditions. The blackchin tilapia (*Sarotherodon melanotheron*) found in Lake Nokoué, a lagoon on the coast of Benin, can tolerate a wide range of water salinity, for example. With warm, tranquil waters, protected from strong currents and from many predators, lagoons are an ideal nursery for juvenile fish, sharks, and

Bora Bora
The islands of Bora Bora, in the south Pacific, are almost completely surrounded by a coral reef, which encloses a lagoon and has just one opening to the ocean.

COASTAL LAGOONS

crustaceans. The rich marine life in turn provides food for many other species, including migrating birds, as well as humans.

Lagoons and people

People around the world have valued lagoons as natural harbors and fishing sites. Most famously, the Italian city of Venice is located on a lagoon and its barrier islands. Human activity, however, can damage the delicate lagoon ecosystem as a result of pollution (including from the tourist industry), land reclamation, and noise from shipping. These low-lying areas, vulnerable to changes in sea level, are also at risk from the effects of climate change, which may alter the level and composition of the water.

FORMATION OF A LAGOON

A coastal lagoon forms along a flat or gently sloping coastal plain, with gentle tides, and a chain of islands, a reef, or a sand bar just off shore. These lagoons develop when the sea level is rising relative to the land, or the coastal land is subsiding. Sea water seeps in and is trapped in a shallow basin with a small inlet to the ocean. Coastal lagoons are subject to continual change from deposition of sediment and organic debris, falls or rises in sea level, and erosion of the sand bar, spit, or reef across the entrance.

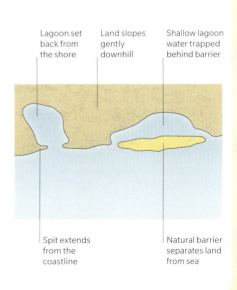

STRUCTURE OF A LAGOON

OCEAN EDGES

Flamingos

Phoenicopteridae

Standing between 2½ ft (0.8 m) and 4 ft 9 in (1.5 m) tall with a distinctive plumage, flamingos are known for their beauty and elegance. However, they are also resilient survivors.

Elongated neck gives the flamingo its distinct silhouette

Coastal lagoons and marshlands with shallow water full of algae and invertebrates are rich feeding grounds for all kinds of birds. In some areas, flamingos that have finished breeding will head to such places and join the foraging flocks of smaller shorebirds. However, in general flamingos are associated with rather more hostile environments. In Africa, lesser flamingos are nicknamed "firebirds," because colonies appear almost overnight, phoenixlike, to breed on hot, open soda lakes. Elsewhere, they breed on salt flats where the water is too saline to support much life. They are extremophiles, able to tolerate conditions that are too harsh for other species. However, they complete their breeding cycle at high speed, in case conditions become too challenging even for them.

Flamingos feed with their heads upside down in the water. Their unique bills, lined with long, hairy fibers called lamellae, filter out edible morsels, especially brine shrimp and algae, whose pigment colors the flamingo feathers pink (see below). A flock will walk synchronously across the water, and this close sociality is also seen in the timing of their breeding, with all pairs building their mud nests and laying eggs at around the same time. Entire colonies are vulnerable to flooding, but the chicks can leave the nests and be shepherded to safer places at just a week old.

Anyone for croquet?
In *Alice's Adventures in Wonderland* (Lewis Carroll, 1865), Alice plays a curious game of croquet using a live—and reluctant—flamingo as a mallet.

" The tongue of the Flamingo is remarkable for its texture, magnitude, and peculiar armature "

Natural History, Or, Second Division of "The English Encyclopedia," Volume 2, 1866

COLOR ORIGINS

Some of the pigments in birds' feathers come directly from the food they eat, including the beta-carotene that makes flamingo feathers pink. For many decades, flamingos kept in zoos were white until ornithologists discovered the need to give them foods extremely rich in this carotenoid pigment. Because their color is diet-derived, wild flamingos naturally show considerable variation in intensity of color at different times of year and states of health.

FLOREANA ISLAND, GALAPAGOS

American birds
The American flamingo is also known as the Caribbean flamingo. Naturalist John James Audubon painted this scene after observing the species in 1832 near Indian Key, Florida.

FLAMINGOS

GREATER FLAMINGO
Phoenicopterus roseus
This flamingo, found in Africa, Asia, and Europe, is the world's largest species. It breeds around salty lakes and coastal lagoons.

AMERICAN FLAMINGO
Phoenicopterus ruber
The most colorful flamingo, this large, mainly coastal species occurs from Mexico to northern South America.

ANDEAN FLAMINGO
Phoenicopterus andinus
This flamingo lives only on high Andean salt lakes. Its habitat is threatened by mining for borax.

Chesil Beach

This geological oddity on the south coast of England has become famous for its unusual characteristics, its abundant fossils, and its rich history and folklore.

At its southern end, **Chesil Beach** could be moving east at a rate of around **6 in (15 cm)** per year

Chesil Beach in Dorset, UK, is a tombolo connecting to a tied island (see pp.82–83). It is unusual for a tombolo because it runs parallel to the shore, like a spit, rather than projecting out into the ocean like a "normal" tombolo. It stretches 18 miles (29 km) along the Jurassic Coast, an area rich in fossilized remains from around 185 million years ago, and is on average 525 ft (160 m) wide. It forms a barrier between the sea and a large coastal lagoon along most of its length, and joins to a tied island, Portland, at its southeastern end. It is believed to have been formed when rising sea levels after the last ice age changed the shape of the coastline by flooding low-lying areas of land and leaving large amounts of sand and pebbles. Waves and currents carried and deposited pebbles to create the beach.

Stones and smugglers

Chesil Beach is also notable for the unusual gradation of its pebbles—around 180 billion of them—which are naturally sorted by size by tidal currents: the small, pea-size ones cluster at the northwestern end and they steadily grow larger toward Portland in the southeast, where they are around the size of a fist. Around the 17th and 18th centuries, when smugglers used to land on Chesil Beach at night, it is said they could tell exactly where they had landed by looking at the size of the pebbles.

During large storms, 3.3 million tons (3 million metric tons) of pebbles can be washed off the beach. Although some are later returned by the tide, they are never fully restored. Cliff erosion used to supply Chesil Beach with more pebbles, but human activities have stopped this and the beach is no longer naturally replenished.

Shipwrecks and tall tales

The beach is renowned for shipwrecks. Strong currents and winds push boats toward the shore, while the rocky seabed makes it hard to anchor, and the steepness of the beach inhibits vessels from landing safely. Hundreds of wrecks were recorded here from the 16th to the 20th centuries, including ships from Spain, Sweden, and North America.

Since the 15th century, people have claimed sightings of the mythical Chesil Beach monster, Veasta. This local legend has been described as a huge rooster, or a monster that was part-man, part-hog, or a giant creature that was half-fish, half-seahorse.

Lagoon border
Chesil Beach forms a natural barrier to Fleet Lagoon, England's largest coastal lagoon (see pp.76–77), home to seagrass, sandworms, fish, sea birds, and more.

Maritime heritage
Seafaring and fishing are an integral part of the history of the local people of Chesil Beach, as seen in this 1899 magazine illustration.

Star-shaped crinoids embedded in a rock from Chesil Beach

Fossil finds
Chesil Beach contains large numbers of fossils from the Jurassic era, including crinoids, ammonites, and belemnites (see pp.62–63).

> " It is above all an elemental place, made of sea, shingle and sky, its dominant sound always that of waves on moving stone. "
>
> JOHN FOWLES, *The French Lieutenant's Woman*, 1969

Tied Islands and Tombolos

Tied islands are islands that are connected to the mainland by a narrow bank known as a tombolo. These curious coastal features are formed by the action of currents.

Estonian islands
This tied island is connected by a tombolo to the north coast of Hiiumaa, Estonia's second-largest island, in the Baltic Sea.

At least half of the islands in the ocean sit near a mainland coastline: these are fragments of continent that have become separated from the mainland. In some cases, this is because of a large fault lifting up blocks of rock that become isolated by erosion. In other cases, the islands are made of a hard resistant rock, such as granite or basalt, that is left behind when softer sediments adjacent to the mainland are eroded. Many islands were formed when sea levels rose by around 400 ft (120 m) at the end of the Last Glacial Maximum, between 18,000 and 8,000 years ago.

These islands are battered by waves, tides, and coastal currents that sweep the shoreline, eroding the land, and carrying sediment across the seafloor. The presence of an island also refracts and diffracts these currents, which can result in a sand or shingle isthmus building up between the island and the mainland or another island. This narrow strip of new land is a tombolo, making the island a tied island. In places where the tombolo rises only just above sea level, it may be exposed and flooded by the fall

Low and high tides at **Mont-St-Michel**, France, vary by up to **46 ft (14 m)**

TIED ISLANDS AND TOMBOLOS

and rise of the tide to create a tidal island. There are many thousands of tied islands around the world. St. Ninian's tombolo, part of the Shetlands archipelago, is a 1,640 ft (500 m) stretch of sand that joins St. Ninian's Isle to Shetland mainland, in northern Scotland, while the islands of Miquelon and Langlade—part of the French overseas collectivity of Saint Pierre and Miquelon, off the coast of Newfoundland, Canada, in the Atlantic—are joined by a 7½ mile (12 km) tombolo.

Beliefs, uses, and inspiration

Because of their charm and mystery, islands that become isolated at high tide sometimes assume religious significance—such as Lindisfarne in Northumberland, UK, and Mont-Saint-Michel in Normandy, France. Others are strategically significant: the city of Hakodate in Japan is built on a tombolo joining Mount Hakodate to the main island of Hokkaido, and was one of Japan's first ports to open to foreign trade. The Rock of Gibraltar is an imposing limestone mountain that has served to guard the western entrance to the Mediterranean Sea for several millennia. It is joined to mainland Spain by a 5½-mile- (9-km-) long tombolo—now the site of an airport runway.

Burgh Island in south Devon, UK, is the setting for two of crime writer Agatha Christie's novels—*Evil Under the Sun* and *And Then There Were None*. The latter features a group of people trapped on the island by the high tide and a murderer picking off victims one by one.

Cormoran the giant features in folklorist Joseph Jacobs' telling of *Jack the Giant Killer* and is killed after Jack lures him into a pit he has dug on St. Michael's Mount

Cornish giant
According to Cornish legend, the tidal island of St. Michael's Mount in Cornwall was built by a mythical 18-ft (5.5-m) giant named Cormoran using stones brought from the mainland with the help of his wife. From his new lair he raided livestock and was eventually killed by a local boy.

Sandy Beaches

A common coastal landform around the world, sandy beaches are formed by low-energy waves depositing sediment on shorelines. They are typically found in shallow, sheltered bays.

A beach profile is the name given to the theoretical cross section of a beach, and sandy beaches typically have gently sloping beach profiles, with ridges marking the high-tide and storm-tide zones. The sand on beaches was once rocks, shells, or corals that have been broken down over time by the elements into ever smaller shards, pebbles, and, eventually, individual grains. As with most types of beach, the material nearest the water is smallest, because it is subject to the most erosion, and the smaller and rounder the grains, the softer the sand feels.

Beach sand can be different colors, depending on its composition. The sands of red beaches, such as those on Rábida Island in the Galápagos, have a high

The first US **public beach, Revere Beach,** Massachusetts, opened in **1896**

SANDY BEACHES   85

iron content. Golden sand is rich in weathered iron (iron oxide), and green sand contains high quantities of the mineral olivine. Pink Sands Beach in the Bahamas gets both its name and its color from the eroded remains of the red shells that were once home to tiny organisms called foraminifera blending in with the sand. The striking black sand of Reynisfjara in Iceland, meanwhile, was once lava.

Pure white sand is made of quartz or coral skeletons ground into dust over millennia. Coral also becomes sand when its skeleton is eaten, digested, and excreted by parrotfish (see p.85). The beach with the whitest sand in the world is Tulum in Mexico, which researchers have assessed as being just 1.4 points off the purest shade of white. Just behind it are Anse Source d'Argent in the Seychelles and Marmari Beach in Greece. Only two beaches in the world—Shell Beach and L'Haridon Bight, both in Australia—are made entirely of shells rather than sand.

Beach life
Many animals have adapted to live on or near the sand. Burrowing species such as lugworms (*Arenicola*), various razor clams, and crabs dig holes in it where they can hide safely from predators. Sea turtles (Chelonioidea) and some ground nesting birds, such as terns (Sternidae) and plovers (Charadriidae), lay their eggs in sand until they are ready to hatch. Groups of baby sea turtles, which are particularly vulnerable to predators, hatch at almost the same time. This is a defense mechanism the species has evolved to lessen each hatchling's chances of being eaten as they race down the beach to the relative safety of the sea (see pp.88–91). Other animals, such as seals, also come onto beaches to rest, digest their food, or give birth.

Human visitors
Sandy beaches are also widely used by humans for leisure activities: in 2023, coastal and marine tourism was estimated to account for more than

> " ... the waves come bounding to the shore, foaming and sparkling, as if wild with glee. "
>
> ANNE BRONTË, *Agnes Grey*, 1847

Rainbow sands
There are around 70 different colors of sand on Rainbow Beach in Queensland, Australia. The varying hues are caused by different minerals in the sand.

Beach cleaners
Painted ghost crabs (*Ocypode gaudichaudii*) search through the sand for food, cleaning the beach and leaving tiny sand balls behind.

Sandy dots on the crab's square shell provide camouflage against predators

50 percent of the global vacation industry. Notable locations can be found worldwide: the world's longest beach is the 157-mile (253-km) Praia do Cassino in Brazil, and the shortest is the 131-ft (40-m) Playa de Gulpiyuri beach in northern Spain. In India, Chandipur Beach is often referred to as the "hide-and-seek" beach, because the ocean recedes so far out during low tide—3 miles (5 km)—that it looks as though the sea has disappeared altogether.

Under threat

While the size, shape, and extent of sandy beaches are in a constant state of change due to tides, waves, and wind, they are becoming increasingly affected by human activity. Climate change has led to a rise in sea levels and an accompanying loss of beaches. Pollution caused by the water runoff from homes and businesses, and the discharge of ships and sewage outlets, also affects the condition and the biodiversity of sandy beaches, as does the human-created garbage that washes up on the shore. While studies show that the coasts of Australia and Africa are most by affected these issues, it is a global phenomenon: at Freeport, in Texas, the beach at one 11-mile (17-km) stretch of coast is disappearing at a rate of 49 ft (15 m) each year. The conservation status of beaches has also been impacted in the last two centuries by the increased use of coastlines for leisure activities.

Shore stories

Kōkī Beach on the Hawaiian island of Maui has dark iron-rich red sand from the nearby cinder cone Ka Iwi o Pele. The name, which means "bones of Pele," comes from a Hawaiian folktale that recounts how the volcano goddess Pele was killed by her sister Namakaokaha'i on that location.

According to First Australian stories, the multicolored sands on Rainbow Beach in Queensland were formed after a woman called Murrawar fell in love with the rainbow but was abducted from her people and forced to marry an abusive man named Burwilla from a rival

> " Oh! I do like to be beside the seaside! I do like to be beside the sea! "
>
> JOHN A. GLOVER-KIND, *music hall song,* 1907

Black markings on back, neck, and head

Long orange-red bills used to prize open or hammer into shells to reach food

Dark wader
Once called sea pies, oystercatchers (*Haematopus ostralegus*) are often found on sandy beaches. They leave distinctive trails in the sand as they look for food such as cockles and mussels.

Lowestoft, Sussex, UK
During the 20th century, sandy beaches became a symbol of leisure and escapism, as seen in this idealized railroad advertising poster.

community. When Murrawar tried to escape, the rainbow came to her rescue. Burwilla attacked it with his boomerang, breaking the rainbow into a thousand pieces and creating the multicolored sands that exist there today.

legend, Blackbeard's spirit is said to haunt the island and its surroundings, with some people reporting seeing a strange light in the water or hearing the ghostly sound of the pirate's head calling for its lost body.

A deadly encounter

A sandy beach that is said to be haunted by a famous figure is Ocracoke Island off the coast of North Carolina. It is where the 18th-century pirate Edward Teach—better known as Blackbeard—is believed to have anchored his ship and where he was ambushed by a force sent by the governor of Virginia, at a spot now known as Teach's Hole. After supposedly shooting Teach five times, the governor's men stabbed him, and then cut off his head and threw him into the sea. In local

PARROTFISH AND SAND

Sand can be created by parrotfish. They graze on hard coral using their tough tooth plates. The coral breaks down further in their guts and is excreted as fine sand. One parrotfish can create around 1,000 lb (450 kg) of sand each year.

PARROTFISH TEETH

Eyes function well both in and out of the water

Large carapace is the biggest of all hard-shelled species

Paddlelike, flexible flippers allow turtles to travel vast distances and move in high-speed bursts

The collective **name** for a group of **turtles** is a **flotilla** or a **bale**

Sea Turtles

Cheloniidae, Dermochelyidae

Found throughout the world's ocean except the polar seas, these sea reptiles have developed a host of distinctive adaptations, from carapaces to navigation abilities.

Sea turtles have existed for more than 100 million years. Many species became extinct, but seven species have survived. They range in size from the Kemp's ridley sea turtle (*Lepidochelys kempii*), at around 2 ft (60 cm) long, to the leatherback (*Dermochelys coriacea*) that can reach up to 6 ft (1.8 m). Six of the seven living species are in the Cheloniidae family, characterized by a hard shell, which helps to both protect and streamline them. The shell, or carapace, is composed of keratin plates called scutes. Each species has a distinct arrangement and number of scutes. The leatherback, the only living species in the Dermochelyidae family, does not have a hard shell, and its back is covered with thousands of bony plates embedded in a layer of fat and oil, with a leathery skin over the top. Partly due to this flexible shell, leatherbacks can cope with colder waters than other species, allowing them to undertake deep dives of around 3,300 ft (1,000 m) and survive closer to the poles.

Unusual diet

Unlike other sea turtle species, green sea turtles (*Chelonia mydas*) are herbivores, feeding primarily on seagrass and seaweed. This diet is thought to give their fat a green tinge, hence their common name.

SEA TURTLES

Because they are reptiles, turtles need to come to the surface to breathe. Despite this, they spend most of their lives underwater. By charging their blood with oxygen and slowing their heart rate down to one beat every nine minutes, they can stay submerged for up to seven hours at a time.

Sea turtles are mostly carnivorous and hydrate by drinking salt water. They have developed the ability to extract surplus salt from the seawater they have ingested and dispose of it via glands beside their eyes. When seen on land, turtles often appear to be crying as they excrete the excess salt. Leatherback turtles have extra-large salt glands—twice the size of their brains—because they feed mainly on jellyfish, which are 95 percent seawater.

Navigation and nesting

Turtles grow very slowly and it takes many years, often decades, for female sea turtles to reach reproductive maturity. When they are ready, many sea turtles migrate thousands of miles between their feeding grounds and their nesting beaches. They make this long journey to return to the same beach on which they were born to lay their own eggs, a process known as natal homing. The species that travels the farthest is the leatherback turtle, which swims up to 12,000 miles (19,000 km) each year.

How female sea turtles find their way back to their nesting beaches after decades away has long fascinated scientists. Research suggests at least two essential components are involved. The first is the turtles' ability to sense the Earth's magnetic field and use it to guide them. The second is their highly developed sense of smell, which may allow them pick up the familiar scent of their birth beach.

When she arrives at the beach, a female waits for nightfall before leaving the water, slowly heaving her body along the sand with her front

Human figure is depicted standing triumphantly on a turtle

Mayan figure
The ancient Maya peoples of Mesoamerica believed that the Earth was a giant turtle floating on a never-ending sea.

In many ancient cultures, sea turtles are believed to possess supernatural powers

Hindu icon
Kurma, meaning turtle or tortoise in Sanskrit, is an avatar or manifestation of the Hindu god Vishnu, one of the religion's main deities. This avatar of Vishnu is often portrayed as a human figure with the lower half as a tortoise or turtle.

Hatching turtle
A baby loggerhead turtle (*Caretta caretta*) breaks its egg open. Sea turtles have an "egg-tooth" on the bottom jaw, called a caruncle.

Egg shells are soft, rather than brittle

flippers. Once she crosses the high-tide line, she chooses an appropriate spot and begins digging, first creating a body pit with her front flippers and then using her back flippers to excavate a deep, narrow hole. She then lays around 100 white, ping-pong-ball-size eggs into this hole before refilling it and flicking loose sand over the nest to camouflage it. Only then does she return to the sea. She will repeat this process multiple times during the nesting season.

New life

Around two months later, usually at night, the young turtles begin to hatch and climb out of the nest. As soon as they emerge, they begin to move toward the water, finding their way by locating the brightest light source, which is usually the moonlight reflecting on the ocean surface.

Only an inch or so in length, turtle hatchlings are highly vulnerable to predators. Birds, crabs, dogs, and raccoons wait near the nests for their prey to appear. Even those hatchlings that make it to the water are far from safe—sharks and other fish hunt in the shallows, preying on the newborn turtles. It is estimated that fewer than 10 percent of turtle hatchlings escape predators in

> " The female turtle lays eggs like those of birds, one hundred in number. "
>
> PLINY THE ELDER, *Natural History,* 77 CE

LEATHERBACK SEA TURTLE
Dermochelys coriacea
The largest living species of sea turtle, leatherbacks can weigh up to around 2,000 lb (900 kg) in adulthood.

FLATBACK SEA TURTLE
Natator depressus
Compared to the high, domed carapace of most turtles, the flatback's shell is flat and smooth. They have a maximum weight of 200 lb (90 kg).

HAWKSBILL SEA TURTLE
Eretmochelys imbricata
This species weighs up to 175 lb (80 kg). It is named after its pointed, beaklike mouth, which it uses to extract sponges from narrow crevices.

their first few days. Once in the open sea, most species of young sea turtles swim to join ocean gyres. They remain in the open ocean for between 5 and 10 years before returning to coastal waters.

Cultural connection

Sea turtles have been revered by many peoples around the world for thousands of years, including the Iroquois people of eastern North America, the Moche people of Peru, and the Seri people from the Gulf of California, whose stories all tell that the world was created on the shell of an enormous sea turtle.

Unfortunately, sea turtles' cultural significance has not been enough to protect them from the impact of human development. Sea turtles have been hunted by humans for millennia—archaeologists have discovered turtle bones and shells in middens dating back 7,000 years. However, human activities have had a particularly devastating effect during the last 200 years.

As well as being overexploited for their meat, eggs, shells, and oil, sea turtles face many other dangers, including becoming accidentally captured in fishing gear; being injured by vessel propellers; and the effects of pollution, habitat destruction, and climate change. However, populations are improving for several species thanks to conservation efforts.

Flood controller

An ancient Chinese legend tells of Yu the Great, who controlled devastating floods in the region with the help of a yellow dragon and a powerful black turtle.

TEMPERATURE AND SEX

In most species, sex is determined during fertilization, but in sea turtles it depends on the temperature of the sand in which the eggs are buried. Eggs incubated below 81.86°F (27.7°C) result in male hatchlings, whereas temperatures above 88.8°F (31°C) produce a clutch of females. Temperatures in between produce a mixture of both. Studies suggest that climate change-induced warming is skewing the ratios, leading to an overabundance of females and a shortage of males. Above a certain threshold, raised temperatures can also reduce the number of hatchlings that survive and even cause entire nests to fail.

EFFECT OF NEST DEPTH

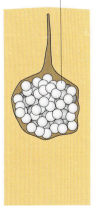

Eggs at the top of the nest are warmer, so more likely to be female

SHALLOW NEST

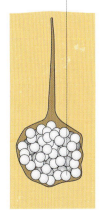

Deeper nests are cooler, so result in more male hatchlings

DEEP NEST

Symbiotic relationship
An olive ridley turtle (*Lepidochelys olivacea*) off the Mexican coast is surrounded by a group of cleaner fish including Yellowtail surgeonfish (*Prionurus punctatus*), Mexican hogfish (*Bodianus diplotaenia*) and King angelfish (*Holacanthus passer*). Various species of fish remove parasites from turtle shells—a symbiotic relationship that provides the fish with food. Olive ridley turtles are known for returning to the same beaches to lay eggs in huge *arribadas* (mass nestings) of thousands of individuals. They lay their eggs in nests around 1½ ft (45 cm) deep that they dig with their back flippers.

OCEAN EDGES

Taking to the air
Pelicans are large birds and need to expend a lot of energy to get airborne. Once aloft, they soar on thermals of warm air with little effort.

Long bill allows for an extra-capacious pouch

Totipalmate feet have webbing linking all four toes

Australian Pelican

Pelecanus conspicillatus

As a bird that carries its own "fishing net," the Australian pelican is a formidable predator of its marine ecosystem. As well as Australia, they are also found in New Guinea and eastern Indonesia.

Pelicans are among the **heaviest flying birds** in the world. Some species reach **33 lb (15 kg)**

With heavy bodies, long bills, and conspicuous pouches, pelicans are striking water birds. Some species live inland while others, such as the Australian pelican, have a more coastal distribution and often fish at sea, diving and bill-dipping while swimming. Another species, the brown pelican (*Pelecanus occidentalis*) of the Americas, also plunge-dives from the air. Like cormorants, Australian pelicans work together to herd shoals of fish into shallower water, making them easier to catch.

Although not the largest species, the Australian pelican is the longest-billed, not just among pelicans but among all birds—the longest bill measured was 20 in (50 cm). Its overall length from bill to tail can reach nearly 6½ ft (2 m). It is widespread in all watery habitats and nests on the ground in colonies on islands. Outside the breeding season, these pelicans often frequent fishing wharves and harbors and, being so visually distinctive, they are a common theme in local legends and folklore.

A marvelous bill
The Australian pelican's pouch is elastic and expandable, able to scoop up more than 3 gallons (13 liters) of water. This is then

Peruvian pelican pot
This double-spouted vessel from the Nazca culture of 7th-century (or earlier) Peru bears the image of a Peruvian pelican (*Pelecanus thagus*).

Spouts are bridged by a handle

Pelican in Her Piety
Edward Burne-Jones's 1880 painting of a "vulning" pelican shows the bird as representative of Christ, saving his "children" with his own blood. Pelicans regurgitate macerated (and therefore bloody) fish for their chicks.

" And to lose the chance to see ... a file of pelicans winging their way homeward across the crimson afterglow of the sunset ... why, the loss is like the loss of a gallery of the masterpieces of the artists of old time. "

THEODORE ROOSEVELT,
Book-Lover's Holidays in the Open, 1916

squeezed out of the sides, leaving any captured fish behind to be swallowed. It is not known if this "fishing net" method of feeding directly inspired human fishers to create nets; however, an Indigenous Australian tale tells of how the pelican Goolay-Yali invented the net, and taught people in coastal communities how to weave their own. Another legendary pelican used his bill to provide people with fire, by creating sparks while pecking at colorful opal pebbles.

According to 2nd-century CE Greek Christian text *The Physiologus*, a pelican mother pecks her own breast open and feeds her chicks with her own blood, or even uses her blood to revive them if they die. This curious self-wounding or "vulning" is represented in numerous artworks, but in reality, Australian pelican parenting is not so attentive. Although they produce broods of two, the younger chick invariably starves or is killed by its sibling. The survivors band together into groups known as pods when they are about 25 days old. They become progressively more active over the following few weeks, and can fly at three months of age.

Coconut Palm

Cocos nucifera

The coconut palm is thought to have originated on the islands of the western Pacific. It is well suited to live on tropical islands, since it can survive sandy, salty conditions and its seeds are dispersed by the ocean.

Carvings depict coconut trees

Delicate decoration
This antique Chinese cup from the Qing Dynasty (1644–1912) has been carved from a coconut shell and lined with pewter.

Coconut palms are large, perennial, flowering plants of the palm tree family (Arecaceae) found in tropical areas around the world. They can grow up to 100 ft (30 m) in height and their fruit can weigh around 2¼ lb (1 kg). Coconuts have evolved to float on sea water—the hollow interior and fibrous middle layer make them light and buoyant, and the outer covering is water resistant. Experiments have shown that coconuts can stay afloat for up to 110 days (and, in theory, travel up to 3,000 miles/ 5,000 km) and still germinate. However, they are more likely to float to nearby islands.

Coconuts and people

Long-distance dispersal, by contrast, involved humans. From around 3000 BCE, peoples such as the Austronesians of southeast Asia carried coconuts across the Indian Ocean to east Africa, and across the Pacific to South America. The coconut palm is sometimes called the "tree of life" because it has so many uses. Coconuts and their water provide food and liquid. Their wood provides timber, their leaves can be used in thatching or woven into baskets, and coir—the fibrous husk of the coconut shell—can be made into items including insulation and rope or fishing nets; it can also be burned as fuel. In addition, the roots are used by traditional healers in Cameroon to create a mouthwash, and in Cambodia to treat dysentery. For these reasons, coconuts were useful to people migrating across the oceans or establishing communities on desert islands.

Coconuts also feature in legends. The Tagalog people of the Philippines believed that the tree grew from the body of one of the first beings to inhabit the world.

> " *E le falala fua le niu, ae falala ona o le matagi* "
> (The coconut tree does not sway by itself; it sways because of the wind.)
>
> TRADITIONAL SAMOAN SAYING

COCONUT FEEDERS

Coconut crabs (*Birgus latro*), are the largest and heaviest terrestrial crab species. Found on Indo-Pacific tropical islands, they live in rock crevices or burrows. Adults cannot swim, but they start their life at sea—females release eggs into the ocean and the larvae float on coconuts or driftwood until the juveniles return to shore. They feed on fruit and seeds as well as coconuts, using their strong claws to open the tough shell.

COCONUT CRAB ON A COCONUT

Features and uses
This illustration from *Köhler's Medicinal Plants* (1883), by German physician Hermann Adolph Köhler, shows the overall form of the coconut palm, and its fruits' shells and flesh.

Taf. II. Cocos nucifera L. — Palmae (Cocoineae)

OCEAN EDGES

Lugworm cast
Blow lugworms leave messy casts—pictured here—while those of black lugworms are smaller and neatly coiled.

Branching red bristles along the lugworm's sides are gills filled with hemoglobin for carrying oxygen around the body

Branching gills
This lugworm illustration from the 19th-century Dutch *Iconographia Zoologica* clearly shows the difference in size between the fat head and the thin tail end of the worm.

Lugworms

Arenicola

The snaking coils of a lugworm's casts are a telltale sign that these worms inhabit a beach. These unassuming animals may be of use in future medicines.

> If a **lugworm** loses part of its **tail**, it can **grow back**

Lugworms are found on sandy or muddy beaches and on the seabed throughout the world. Living in U- or J-shaped burrows under the sand, they eat tiny organisms in the sand and their feces—known as a "cast"—can be found piled on the surface.

The two most studied lugworm species are found in the European intertidal zone. The blow lugworm (*Arenicola marina*) usually lives higher up the beach than the prized black lugworm (*Arenicola defodiens*). The latter makes a popular bait but can only be found when the tide has gone out. They were recognized as two genetically separate species in 1993.

Special blood
Lugworms live in a low-oxygen environment. As a result their hemoglobin—the protein in blood that carries oxygen around the body—has evolved highly effective oxygen-carrying properties. This may have significant potential for medical use.

Marram Grass
Ammophila

Also known as beach grass, marram grass is a familiar sight on North Atlantic coastal sand dunes. It plays an important role in securing the dunes against the strong winds so other plants can settle and grow.

The thin, spiky leaves of the two species of marram grass—European (*Ammophila arenaria*) and North American (*Ammophila breviligulata*), often seen growing on top of sand dunes—can grow to around 3 ft (1 m) tall.

Withstanding blustery days
Below the surface, long, matted roots help to stabilize the dunes and prevent sand from being swept away in strong winds. Unlike many other plants, marram grass thrives in arid, windy, and salty conditions with low nutrient levels. Each blade of marram grass curls in on itself to trap humid air inside and help the plant retain moisture. These rolled leaves also protect the stomata—tiny openings which control the exchange of gases and water between the plant and its external environment—from strong gusts of wind. Marram grass also has many uses for local communities, and people have historically used it to make thatching for roofs, rope, fishing nets, paper, and shoes, as well as cattle feed.

Golden flowers
Marram grass can be seen year-round, but it flowers in July and August before spreading its seed in August and September.

Amongst the Bents
This 1872 oil painting by the Scottish artist William McTaggart shows a group of children hiding and playing among the marram grass.

Dunes

Sweeping banks of sand created by the wind, dunes are a common feature of marine coastlines. Only the hardiest species can survive in this harsh, ever-changing environment.

Short-lived lagoons
At Lençóis Maranhenses National Park in Brazil, storms during the rainy season create temporary freshwater lagoons between the white sandy dunes.

Sand dunes are created when wind blows grains of sand and causes them to accumulate in small mounds. Over years, more sand accumulates and the mounds grow bigger until they become the enormous sandy hills that are often seen on sandy coastlines. Some dunes are thousands of years old, and can grow to great heights. Mount Tempest, on Australia's Moreton Island, is considered the tallest coastal sand dune in the world, reaching 935 ft (285 m) above sea level.

Every dune has a windward side—where the wind blows and adds more sand—and a steeper slip face on the other side, which is protected from the wind. Because they are made of sand, dunes can sometimes collapse unexpectedly. Medieval trader and explorer Marco Polo described an eerie "singing" sound booming across the sand dunes. This was made by avalanches: as the grains of sand plummet down the dune and rub against each other, it creates a loud rumble that can be heard for miles.

Dots and sand
Abstract art pioneer Piet Mondrian captured the vast sand dunes that accumulate along the shores of Zeeland in the Netherlands in this 1909 pointillist oil painting.

" Just as the sand dunes, heaped one upon another, hide each the first, so in life the former deeds are quickly hidden by those that follow after. "

MARCUS AURELIUS, *Meditations*, c. 121–180 CE

Buffers and burials

As they build up, dunes protect the coastline from the elements, buffering strong winds, and preventing flooding. Once established, they encourage plants such as marram grass (*Ammophila arenaria*, see p.99), clovers, seaside sandplant (*Honckenya peploides*), and sea spurge (*Euphorbia paralias*) to grow. Because sand cannot retain water easily, these plants tend to have long roots and a tough outer layer to help them survive in the harsh, dry conditions. The plant life creates a habitat for insects, which attract small mammals, which, in turn, can draw in birds of prey, including kestrels and owls. Ground-nesting birds such as ringed plovers (*Charadrius hiaticula*) and terns (see pp.166–167) may also make their homes on dunes.

In Australia, Indigenous peoples historically held ceremonies to bury their dead in coastal sand dunes. Today, development threatens dunes and their wildlife by causing pollution, vegetation loss, and erosion, and by introducing invasive species.

Thick leaves help sea spurge retain moisture so it can thrive in dry environments

Flowers lack petals or sepals. Instead, they have yellow-green modified leaves called bracts that cup the flower

Dune blooms
Sea spurge is a wildflower native to coastal sand dunes in Europe, west Asia, and north Africa, and an invasive species in Australia.

Seagrasses

Posidoniaceae, Zosteraceae, Hydrocharitaceae, Cymodoceaceae

Growing in clear, shallow seas, seagrasses are the basis of a whole ecosystem, supporting a rich range of fauna and providing a strong environmental buffer.

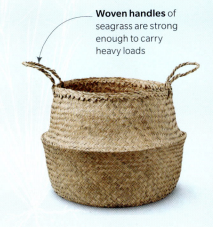

Woven handles of seagrass are strong enough to carry heavy loads

Sustainable seagrass
Seagrass has a long history as a weaving material. It grows quickly and can be sustainably harvested, dried, and used to make baskets, rugs, and cord.

There are around 70 species of seagrass growing in most of the world's seas, mainly around coasts where the water is shallow enough for sunlight to reach the seabed. Seagrasses are flowering plants, just like land grass. They evolved around 100 million years ago from terrestrial plants that moved into marine habitats. Like land grass, they are photosynthetic, and they often grow in meadows that cover large areas of seabed. A seagrass meadow is a living habitat that supports an entire ecosystem.

when blades of seagrass break off, they sink to become food for organisms on the seabed as they decompose.

Environmental hero

Seagrasses also benefit the environment. Their roots stabilize the seabed, holding sand and mud in place, while their blades absorb wave energy, protecting coasts from erosion. They also lock away more carbon per unit area than a rainforest. Like other plants, they help oxygenate seawater.

COMMON EELGRASS
Zostera marina
Common around the north Atlantic and north Pacific, eelgrass gets its name from its slender leaves.

SURF GRASS
Phyllospadix
This genus of seagrass grows in the intertidal zone, and is exposed to the air at low tide.

" All too often the ocean and life beneath the surface is overlooked, but ocean species, like seagrass, truly can help save the world "

SANDY LUK, Marine Conservation Society

Many species feed on seagrasses, including fish, turtles, manatees, crabs, and shrimp. Sea slugs eat microscopic algae that grow on seagrass leaves. The grass provides shelter, hiding places, and nursery habitats for many animals. Various species of fish, including cod and plaice, lay their eggs in seagrass so their young are safely hidden when they hatch. Seahorses, pipefish, and invertebrates such as sea anemones, sea slugs, and snails spend their whole lives among seagrass. Marine worms and mollusks stay safe around seagrass roots and

Moray hideout
A moray eel hides beneath the fronds of a seagrass meadow off the coast of the Italian island of Capri, in the Mediterranean Sea.

OLDEST LIVING ORGANISM

Neptune grass (*Posidonia oceanica*) is a seagrass endemic to the Mediterranean. Scientists have discovered that a patch near Formentera, one of the Balearic Islands, is 80,000–200,000 years old, making it the oldest known living organism on Earth. All the blades of a huge area of Neptune grass 9 miles (15 km) across were found to have the same DNA, showing they had all grown from the same individual. Calculations comparing the patch's size to its growth rate revealed its age.

Skeleton Coast

Along the coastline of northern Namibia, next to the Atlantic Ocean, lies the desolate Skeleton Coast. It is one of the world's wildest places, and true to its name, it is scattered with skeletons of many kinds.

Ocean meets desert
On this typical stretch of the northern Skeleton Coast, constant, crashing waves meet endlessly shifting sand dunes. Often, the only water available to desert wildlife is the sea fog that creeps ashore.

The Namib Desert in southwest Africa is one of the oldest deserts in the world, thought to have formed at least 55 million years ago. More than 20,000 years ago, the local San people began living there, and it is said that they named the Skeleton Coast "the land God created in anger," because of its harsh climate and desolate landscape.

The wind along the coast blows mainly offshore, resulting in very few rain clouds over the land. Conditions are arid and windy, and this means there is barely any vegetation. Sea temperatures are cold due to the Benguela Current from the cooler south, which brings an upwelling of water from the ocean depths. This wild and treacherous stretch of the Atlantic is renowned

STUCK IN THE SAND

The Skeleton Coast is famous for its many shipwrecks, which are now tourist attractions. Many of the wrecks are now some way inland, as the windblown dunes have reshaped the shoreline. Ships such as the *Eduard Bohlen*, which ran aground in 1909, have been half-buried in the sand. Before being left to rust away, this ship was used as accommodation for diamond miners working nearby.

WRECK OF THE *EDUARD BOHLEN*

> **Raw diamonds** are sometimes found on the **shore**, carried along the **Orange River**

for its powerful winds, violent waves, dangerous shifting sandbars, and hidden rocks. Thick sea fogs often hang heavy in the air.

The cold upwelling of deep sea water brings with it nutrients that help feed vast swathes of plankton, which in turn provide a rich food source for marine creatures. The plankton attracts fish, whales, dolphins, turtles, and seabirds such as the Cape cormorant and the greater flamingo.

Cape Cross Seal Reserve, at the southern end of the Skeleton Coast, is home to one of the world's largest colonies of Cape fur seals—more than 100,000 of them. Farther north, around Torra Bay, there is a much rarer coastal creature—the lion. Lionesses can sometimes be seen hunting seals and seabirds on the beaches, especially when droughts make it hard for them to find their natural prey inland. Animal bones litter the shore: remains of oryxes and other desert wildlife killed by lions can be found alongside skeletons of seals and whales left over from the days of whaling.

Skeleton graveyard

Human skeletons have also been found along the coast—the remains of shipwrecked seafarers—and the wrecks of their ships form another kind of skeletal remnant. Since Portuguese trading ships began exploring this coast in the 1400s, dangerous waters and weather have claimed hundreds of vessels and many more lives. The oldest documented wreck is that of the *Bom Jesus*, a Portuguese galleon that sank in 1533. Even those sailors who survived a storm or a shipwreck and made it ashore faced almost certain death, trapped between the inhospitable desert and the ocean, whose powerful waves made it almost impossible to set sail again.

One famous shipwreck story dates from the 1940s, when 12 human skeletons without their heads were found, along with a slate dated 1860. It was left by a survivor, indicating that someone had headed north to find help. But nothing was ever heard of them again, and the mystery of the skeletons' identities has never been solved.

Cape Cross seals
The Cape Cross Seal Reserve on the Namib coast is home to thousands of seals, who feed on the abundant fish from the South Atlantic's Benguela Current.

OCEAN EDGES

Strong swimmers
Sea lions can swim great distances, stay underwater for 20 minutes at time, and dive to more than 1,400 ft (400 m) in search of food including fish, squid and, occasionally, even seal pups.

Sea Lions

Otariidae

Like their close relatives, the seals and walruses, sea lions are marine mammals that spend most of their lives at sea. They return to land to rest, warm up, breed, and escape predators.

Back flippers are rotated underneath the heavy body, allowing the sea lion to waddle on all fours

They may appear clumsy on land, but sea lions are powerful and agile underwater. Their sleek, torpedo-shaped bodies are propelled by their strong, long front flippers, while their back flippers steer them through the water at speeds of up to 30 mph (48 kph). These back flippers mean that, along with seals and walruses, sea lions are known as *pinnipedia*, a Latin word meaning "fin footed."

Sea lions live in subarctic to tropical waters, with the exception of the North Atlantic. Their main predators are killer whales and sharks, but sea lions are formidable hunters themselves. They have specialized eyes that allow them to concentrate light, giving them excellent vision in the ocean depths. When they are closer to the surface, their pupils contract to protect their retinas from strong light. Their highly sensitive whiskers, called vibrissae, detect the slightest movements in the water to help them catch passing prey, which they usually swallow whole.

Mistaken identity
Sea lions are often mistaken for seals. They can be distinguished from each other by both sound (sea lions are much noisier, while seals

SEA LIONS 107

SOUTH AMERICAN SEA LION
Otaria byronia
This large sea lion species is widely distributed across South America. Males can be three times heavier than females with large heads and muscular necks.

NEW ZEALAND SEA LION
Phocarctos hookeri
This is one of the rarest species of sea lion with around 12,000 individuals, found only in New Zealand. Males can reach 11½ ft (3.5 m) in length.

communicate in softer grunts) and by the external ear flaps protruding from the sides of their heads—sea lions' scientific name even means "little ear."

Sea lions are highly vocal animals both below water and above it. They pile onto sandy beaches or rocky shorelines in herds of up to 1,500 individuals, grunting, growling, barking, and roaring at each other. They can be especially raucous during the breeding season, as males compete with each other for territories and females. Pups can swim from birth and walk at just 30 minutes old. After weaning, they may stay with their mothers for up to a year.

Until recently, sea lions were hunted for meat, skin, oil, and even their whiskers, which were used as pipe cleaners. Today they are a protected species, but are often a source of trouble for fishers, competing for fish and becoming entangled in nets.

Small ear flaps with a downward opening

Life on land
Rather than crawling along on their bellies like seals, sea lions can walk and even climb on land. Their strong front flippers support their weight and the back flippers function as feet.

A thick layer of blubber beneath the skin retains warmth and acts as an important energy reserve

Magnificent mane
This copperplate engraving depicts a male sea lion, displaying the lionlike mane that inspired its name.

Exploring the mud
A proboscis monkey (*Nasalis larvatus*) explores estuary mudflats at low tide in Bako National Park, Malaysia.

Islands in a delta
Tidal channels, creeks, and rivulets separate islands of salt marsh (see pp.122–123) in the Bay of Cádiz delta in southwestern Spain.

Estuaries and Deltas

When large rivers reach the sea they often form one of two main landscapes. Estuaries are wide bodies of water, often characterized by strong tidal currents and eddies. Deltas are usually more tranquil, with numerous slow-moving and gently flowing channels.

Estuaries are more likely to form where both the river and sea it empties into are deep, or there is a wide tidal range, sweeping water in and out at high speed. On a rising tide, sea water flows inland and upstream; at low tide, some estuaries almost drain out, revealing an expanse of sand or mud.

A delta, by contrast, is a coastal landform made up of islands of silt deposited by a slow-moving and sediment-heavy river, often emptying from low-lying land into a shallow or sheltered bay. Tides flow in and out of creeks and channels between the islands.

In both environments, this tidal exchange creates mudflats, salt marshes, sheltered shallows, and marshy or forested islands and banksides that encourage a variety of wildlife that can include oysters (Ostreida), bull sharks (*Carcharhinus leucas*), mudskippers (Oxudercinae), and species of crocodiles. Birds such as herons (Ardeidae), terns (see pp.166–167), and some species of storks are also at home in estuaries and deltas.

Each time the tide rises, seawater flows upriver in an estuary, or floods the waterways between islands and raises the river level in a

Waves of enjoyment
Surfers ride the "Mascaret" tidal bore in the Gironde estuary, in France. It occurs between June and October on the Dordogne and Garonne rivers.

1.9 miles (3 km) wide. Moving at up to 25 mph (40 kph), it has been known to overflow the river's banks and wash away unwary spectators.

Human settlements

Deltas and estuaries have been important to humans since prehistoric times. They provide a safe harbor for boats and ships, a sheltered place to catch fish and seafood, and a route inland for travel and trading. Many of the world's largest cities have grown up around estuaries, including Tokyo, New York, San Francisco, London, Hong Kong, and Kolkata. Deltas, being low lying and prone to flooding, are risky places in which to build cities. Even so, conurbations such as Alexandria in Egypt; New Orleans, Louisiana; and Guangzhou and Hong Kong in China have

> " This great estuary is wide, endless.
> The river is brackish, blue with the cold. "
>
> JAMES SALTER, *Light Years*, 1975

delta. This creates a semi-saline blend of salt water and fresh water known as brackish water, to which some species of fish and other marine animals are adapted. These include tule perch (*Hysterocarpus traskii*), rainwater killifish (*Lucania parva*), and estuarine isopods, a type of crustacean. Shellfish, small fish, and turtles shelter in narrow creeks, and silt-enriched estuary mud is full of burrowing worms that attract waders and shorebirds.

Racing tides

In some estuaries, unusually high tides often result in a wave or series of waves rolling upriver in a tidal bore. In a funnel-shaped estuary, the narrowing channel concentrates the wave's energy, focusing its power. The tidal bore on the Qiantang River in Hangzhou, China, is the largest in the world. Its waves can be 30 ft (9 m) high across a stretch of water

grown around the Nile, Mississippi, and Pearl River deltas respectively. This is in part because the mineral rich silt deposited in them is very fertile. The islands of the Ganges delta in India and Bangladesh are dotted with rice paddies and artificially-created ponds used for farming shrimp.

In the event of a major flood, human settlements in these areas are additionally at risk because of the structure of deltas.

Dots represent islands in the delta

Medieval map
This image is a later copy of a map of the Nile Delta made by the 10th-century Muslim explorer and mapmaker al-Istakhri.

ESTUARIES AND DELTAS   111

Their islands, small streams and channels, marshes, and vegetation such as mangrove swamps act as natural barriers that, in normal circumstances, absorb wave energy and protect settlements and ecosystems inland. However, when subjected to major flooding, as happened in the Mississippi delta during Hurricane Katrina in 2005, the results can be catastrophic.

FORMATION OF A DELTA

A delta forms when a flowing river slows down as it reaches the sea and deposits the rock, mud, and silt it is carrying. Heavier sediment forms the delta plain, a flat area just above water cut through with river channels. Lighter sediment is carried farther out to sea, accumulating underwater on the shallow seabed.

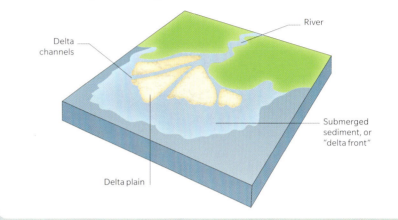

Creating new land
This satellite image of part of the Yukon Delta in western Alaska shows how silt deposits gradually help form new land extending outward into the sea.

Dugong and Manatees

Sirenia

The manatees and the dugong together form a small and fascinating order of mammals—the sirenians, or sea cows. Less well-known than other sea mammals, their closest living relatives are not whales, seals, or walruses, but elephants.

Displays of affection
Manatees, such as these two West Indian manatees (*Trichechus manatus*), have been observed nuzzling and even hugging each other.

DUGONG AND MANATEES 113

Very thick blubber made it hard for Steller's sea cow to stay underwater for long

Rough skin along its sides earned it the nickname "bark animal"

Whalelike tail, like the dugong

Christopher Columbus reported seeing **mermaids,** now thought to be **manatees**

Extinct species
Steller's sea cow was a large species of sirenian that reached up to 33 ft (10 m) in length and was a close relative of the dugong. It was hunted to extinction for food in the 18th century.

There are four species of sirenians: the large West Indian, African, and Amazonian manatees, and the slightly smaller dugong of the Indian Ocean, Australia, and the western Pacific. While the dugong prefers warm, shallow coastal ocean habitats, manatees also swim inland up rivers, especially the Amazonian manatee, which lives mainly in the rivers of the Amazon basin.

Sirenians have no hind limbs, using their tails to propel themselves forward at a leisurely pace. In manatees, the tail is fan-shaped, while the dugong has a two-pointed fluke, like a dolphin's tail. Another difference can be seen in the shape of their heads. Dugongs have a trunklike, flat-ended snout for grazing along the seabed, and males have two small tusks. Manatees have a more rounded, seallike face, allowing them to eat plants at the water's surface too. Both have large nostrils on the tops of their snouts, which can be closed by nasal valves while underwater.

All sirenians are heavy, slow-moving herbivores, although they may occasionally swallow passing shrimp or fish. They spend most of their time grazing—especially on seagrass meadows—earning them their "sea cows" nickname.

Creatures of myth
When not feeding, sirenians sometimes bob upright in shallow water, with their heads above the surface. This may explain why they have been believed to be mermaids—especially manatees, which have a more humanlike facial appearance than the dugong. The name "sirenian" comes from "siren," a part-human creature of Greek mythology, and later a name used for mermaids.

Manatees and mermaids
This 1889 newspaper illustration comparing a manatee with a mermaid was captioned "Real and ideal," comparing the mermaid of fantasy with the supposedly mermaid-like sea creature.

AFRICAN MANATEE
Trichechus senegalensis
This nocturnal manatee makes a whistling sound. It measures up to 15 ft (4.5 m) long and is found in west Africa.

AMAZONIAN MANATEE
Trichechus inunguis
The smallest Sirenian at up to 9 ft (2.8 m) long, this manatee, found in Brazilian waters, has no nails on its flippers.

DUGONG
Dugong dugon
Up to 8–13 ft (2.4–4 m) long, it lives in the Indian and Pacific Oceans. Unlike manatees, it has a two-pointed tail.

Nordic cruise
Norway's fjords have been a popular cruise destination for more than a century. This 1935 advertising poster features the work of German artist Albert Fuss.

Fjords

A fjord, or fiord, is a long, narrow sea inlet. Fjords can reach far inland, and some have several twists, turns, or sub-branches. They were carved by slow-moving glaciers, and filled with water when the glaciers melted and sea levels rose.

Bright red feet make this bird easy to identify

Pigeon guillemot
This striking auk species (*Cepphus columba*) is found close to shore in the Kenai Fjords south of Anchorage, Alaska.

"Fjord" is a Norwegian word, and Norway is home to the world's best-known fjords—more than 1,700 of them, mostly along its western coast. Fjords are also found in other regions that were covered by glaciers during past ice ages, including parts of New Zealand, Scotland, Chile, Greenland, Iceland, and Canada. One of the biggest and longest fjords in the world is Scoresby Sound in Greenland, a branching system of channels that reaches more than 200 miles (320 km) inland.

Fjords typically have high, steep, sometimes almost vertical sides, and very deep water. For humans and other animals, they provide shelter from ocean storms and waves, creating natural harbors. Migrating fish, such as herring, salmon, and mackerel, can be found in fjords at different times in the year.

Deep valleys

As a glacier flows downhill, its force and weight carve out a deep valley, pushing rock and debris ahead of it. This accumulates in the mouth of a fjord as an undersea ridge or "sill." Because of this, a typical fjord is deeper inland than where it meets the open sea. In these inner fjord waters, coral reef ecosystems have been discovered, cut off from the rest of the sea. Unlike tropical reefs, they are adapted to surviving in cold temperatures and with little light.

> " Gudrun went to the sea, and leapt into the water ... she was carried across the fjord, and came to the land of King Jonakr. "
>
> SNORRI STURLUSON, *The Prose Edda*, 1220

DUSKY SOUND

This notable fjord lies on the southwestern tip of South Island, New Zealand. It is New Zealand's longest fjord at 25 miles (40 km), and its rich marine ecosystem is home to black coral and kelp forests, as well as seals, whales, dolphins, and the Fiordland penguin (*Eudyptes pachyrhynchus*). It was used as a harbor by European explorers in the 18th and 19th centuries.

ERRINA DENDYI HYDROZOA IN DUSKY SOUND

Beluga Whale

Delphinapterus leucas

Found around the fringes of the Arctic Ocean, the name of this sociable, intelligent, and famously photogenic whale comes from the Russian word for white (*belukha*). It is also known as the "sea canary" or "melonhead."

For a whale, the beluga is small, averaging 13 ft (4 m) in length—comparable to a large dolphin. Like several other whale and dolphin species, belugas are often found in groups or "pods," usually of around 10–20 individuals, which sometimes form much bigger aggregations of hundreds or even thousands of whales. A pod can be made up of family members or close-knit individuals. They hunt as a team, herding fish into shallow water to make them easier to catch. They also play together, chasing, breaching, and diving in unison, and even hugging each other. Blowing bubbles appears to be another form of play, and belugas have been seen picking up and carrying objects they find, and copying each other's calls.

Arctic species
A beluga whale swims under ice near the Arctic Circle in western Russia. They are also found in waters off Alaska, Canada, and Greenland.

White whale art
This hand-colored steel engraving by William Lizars appeared in The Naturalist's Library, a series of 40 volumes edited by Scottish naturalist Sir William Jardine.

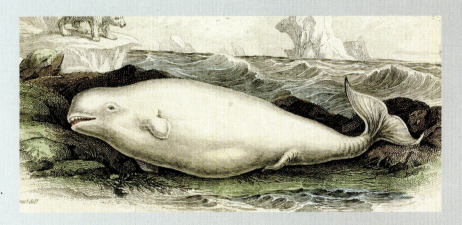

Belugas are known as "sea canaries" because of the remarkable array of sounds they make—squeaks, moos, chirrups, trills, and whistles. They use these sounds for communication, signaling to each other as they travel and interact. Scientists studying these vocalizations

> " They are not often shy, but often follow the ships, and tumble about the boats in herds of thirty or forty; bespangling the surface with their splendid whiteness. "
>
> **WILLIAM JARDINE,** *The Naturalist's Library, Vol. VI: On the Ordinary Cetacea or Whales,* **1837**

have discovered that the whales use different sounds in different situations, and that different regions and pods have their own "dialects." Belugas use echolocation to locate prey, and to find breathing holes when under the ice. They have a large, fatty lump on their foreheads, called a melon, and they use this to focus and direct a stream of clicking sounds, which bounce off objects and echo back to the whale to pinpoint prey. Since they are not fast swimmers, beluga prefer to forage on the seabed or in shallow water for fish, squid, octopuses, shrimp, and worms.

Surviving the ice
The beluga is one of the most northern-dwelling of all whales, rarely straying south of the Arctic Circle. To keep warm in the freezing conditions, they have a thick layer of fatty blubber under their skin, which can make up 40 percent of a beluga's total weight. As another adaptation to its icy habitat, the beluga has no dorsal fin, making it easier to swim smoothly under the sea ice. However, many pods spend only part of the year in icy waters. During summer, when ice cover in the Arctic Ocean shrinks, they move to the coasts and estuaries of Canada, Europe, and Russia to breed, sometimes even swimming up rivers. These whales are highly curious. The US and Russian navies have trained belugas to carry equipment and find objects underwater. Belugas have also been known to rescue divers, find and return possessions dropped from boats, and even imitate human speech.

Social animals
Photographed from above, a large pod of beluga whales gathers in the Canadian Arctic Region.

Kelp Forests

Thick, lush forests grow beneath the sea. Instead of trees, these marine forests, which grow in shallow coastal waters, consist of collections of tall, dense kelp seaweed.

Although it is green (or brownish) and photosynthetic, kelp is not part of the plant kingdom, but is a type of multicellular algae. There are dozens of species of kelp growing worldwide, with the biggest, giant kelp (*Macrocystis pyrifera*), growing more than 160 ft (50 m) in height—taller than many forest trees on land.

Hanging on
To survive in the ever-moving ocean, kelp clings to the rocky seabed using strong holdfasts. These abundant kelp stems, or "stipes," and flattened fronds absorb wave action, calming the sea near the coast. Some species, including giant kelp, have gas bladders that help float the seaweed up toward the surface, so they can reach up toward the sunlight.

Kelp is an important part of the ocean ecosystem. It grows much faster than land forests—giant kelp can grow more than 20 in (50 cm) in a day—taking up large amounts of carbon from the atmosphere. Kelp forests are also vital ecosystem "producers," forming the basis of a coastal

Fronds of kelp with blades to capture light

Giant kelp
In giant kelp, gas bladders, or pneumatocysts, grow at the base of each flattened blade. The kelp fills them with waste gases.

Safety in the fronds
A school of blacksmith fish (*Chromis punctipinnis*), and a small kelp bass (*Paralabrax clathratus*), swim through fronds of a giant kelp forest in the Channel Islands off California.

" The number of living creatures of all Orders, whose existence intimately depends on the kelp, is wonderful. "

CHARLES DARWIN, *The Voyage of the Beagle,* 1839

marine food web. They feed sea urchins, snails, abalones, crabs, and fish. In turn, bigger fish, lobsters, octopuses, starfish, sea lions, and sea otters feed on the kelp-eaters. Kelp provides a place to hide from predators and shelter from rough seas and storms.

Humans can eat kelp—it is a rich source of vitamins, trace elements, and fiber. It can be used to make noodles or sushi, and could become an important human food source in the future.

Balancing the ecosystem
Starfish such as the giant-spined star (*Pisaster giganteus*) feed on sea urchins and other kelp-eating invertebrates. This balances the ecosystem and preserves the kelp.

Floating raft
A group of sea otters floating together is known as a "raft." Sometimes they may even hold each other's paws.

Sea Otter

Enhydra lutris

A unique marine mammal, the sea otter is the only member of the otter family to fully live in the sea. It is found only in the northern Pacific Ocean, from Japan and Russia to North America, as far south as Mexico.

Although sea otters have four legs and can walk, they can spend their whole lives at sea. However, despite mainly living in cold northern waters, they are the only sea mammals without blubber, the thick layer of fat under the skin that keeps animals such as seals warm. Instead, sea otters have the densest fur of any animal, with up to one million hairs per sq in (150,000 per sq cm). Longer guard hairs keep the water out, while shorter, softer fur underneath traps warm air next to the otter's skin. Because of this layer of trapped air, and their large lungs, sea otters float easily. When they are not diving below the surface to find food, such as crabs, sea urchins, and shellfish, they spend much of the day lying on their backs on the surface. A sea otter will hold its catch on its stomach, using a rock as a tool to crack or open the hard shells to reach the edible flesh. The rest of the day is spent napping or grooming. At night, they sleep while floating, often in groups of up to 100 otters.

Mothers and pups
Sea otters are devoted mothers—a female gives birth to a single pup, and spends hours cleaning and nursing it while holding it on her stomach, away from the cold water. As the pup grows, the mother will leave it on the surface, sometimes tangled up in seaweed so it cannot float away while she goes hunting. She teaches the pup to dive for food, starting by collecting pebbles or starfish, before it learns to catch mussels, abalones, and other prey.

Head is carved to resemble the bow of a kayak

Walrus ivory carving
This hat fastener was made around 1800 by the Aleut people. It is carved into the shape of a female sea otter, with her pup resting on her stomach.

Fur color
Sea otter fur is dark brown, but the heads and faces are paler, ranging from reddish-brown to creamy white.

Sea otters have folds of fur under their forelimbs that they use as pockets to hold food or rocks

Hungry visitors
American white pelicans (*Pelecanus erythrorhynchos*) and dunlins (*Calidris alpina*) flock together on a salt marsh in Morro Bay, California, where they forage for fish, insects, and worms.

Salt Marshes and Mudflats

In estuaries, bays, or along sheltered coastlines where the tide rises and falls, mudflats can form, and over time can develop into salt marshes. These important wetland habitats are home to a variety of marine wildlife.

The mudflat and salt-marsh ecosystem changes and develops over time. In areas sheltered from strong waves, a layer of mud gradually collects either as a result of sediment being washed ashore or deposited by rivers as they meet the sea. A typical mudflat is covered by the sea at high tide, and exposed twice a day as the sea retreats.

As more mud collects, it rises higher, creating some areas that remain above the surface even at high tide. Specially adapted, salt-tolerant grasses and other plants, such as marsh samphire (*Salicornia europaea*) and cord grass (*Spartina*), are then able to colonize these mud islands. These establish a network of roots that hold the mud together and trap more and more sediment, building the mud into permanent marshy islands. When old salt-marsh plants die and decay into the soil, they contribute more organic matter, creating a layer of carbon-rich peat over the mud. Unlike many delta islands (see pp.108–111), salt marshes are wetlands. Even though the surface of the salt marsh and its plants may rise above the high-water level, the marsh is still flooded by saltwater, which drains back out through a network of tiny creeks and ditches

Early colonizer
Marsh samphire is one of the first plants to colonize mudflats, and is a traditional wild food.

Fleshy leaves have a salty flavor

> **Salt marshes** are among the most **biologically rich habitats** on Earth

SALT MARSHES AND MUDFLATS

where small fish, crabs, terrapins, and eels can shelter. Though efficient as a carbon sink, the permanently waterlogged mud and peat can release sulfurous gases, giving some mudflats and salt marshes a distinctive smell.

Unique ecosystem

The mud, silt, and marshy soil are full of nutrients and bacteria, providing a habitat for worms, shrimp, snails, and cockles that burrow in the mud, and for insects and other small arthropods that thrive on salt-marsh islands. In turn, these attract hundreds of species of birds, including herons and egrets, pelicans, dunlins, oystercatchers, and—only on the Atlantic coast of North America—the saltmarsh sparrow (*Ammospiza caudacuta*). Some forage on the mudflats at low tide, while others hunt invertebrates on the salt-marsh islands, or use them as a place to fish from when the water is high.

Salt-marsh habitats also support a huge variety of plants, each with its own level of salt tolerance, growing at different heights and locations on the marsh, from samphire and sea purslane to bright-flowering yellow marsh jaumea (*Jaumea carnosa*), pink saltmarsh mallow (*Kosteletzkya pentacarpos*), and purple sea aster (*Tripolium pannonicum*) and sea lavender (*Limonium vulgare*). In turn, these flowers attract flying insects. In some parts of the world, sheep and cattle are let on to salt marshes to graze, benefiting from their nutrient-rich grasses.

Marsh tale

An illustration for the 1917 poem "Overheard on a Saltmarsh" depicts fanciful marsh inhabitants—a goblin and a nymph in conversation.

Saltwater wetlands

Tiaozini Wetlands in Jiangsu province, China, is part of the Yellow Sea-Bohai Gulf Migratory Bird Habitat, and a world heritage site. The wetlands experience a dramatic color change in the fall when *Artemisia halodendron*, a common shrub in the area, turns red. Each winter, tens of thousands of migratory waterbirds flock here in search of food and rest, including critically endangered species such as the spoon-billed sandpiper (*Calidris pygmaea*) and the spotted greenshank (*Tringa guttifer*).

OCEAN EDGES

A DIVERSE TOOLBOX

Waders vary in bill size and shape, and accordingly in their preferred diet and feeding method. This is an example of niche separation, and means that many different species can forage in the same area without excessive competition. They can all benefit from feeding in large mixed flocks because more eyes mean that there is a greater chance of spotting danger, and flying up en masse can confuse a bird of prey.

WADERS' BILLS

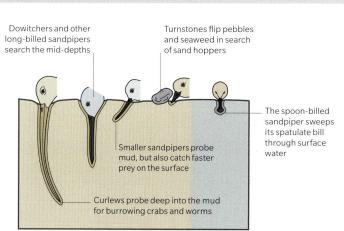

Dowitchers and other long-billed sandpipers search the mid-depths

Turnstones flip pebbles and seaweed in search of sand hoppers

The spoon-billed sandpiper sweeps its spatulate bill through surface water

Smaller sandpipers probe mud, but also catch faster prey on the surface

Curlews probe deep into the mud for burrowing crabs and worms

Swamp hunter
The tricolored heron (*Egretta tricolor*), here painted by John James Audubon in 1834 for his famous *Birds of America*, is mostly found on well-vegetated coastal marshes, where it preys on fish and other aquatic creatures.

The nonstop flier
Bar-tailed godwits (*Limosa lapponica*) breeding in Alaska undertake a record-breaking southward migration over open sea to Australasia. One tracked individual covered more than 8,390 miles (13,500 km) in 11 days, without a moment's rest.

Pectoral muscles increase in size to prepare for migration, and the bird stores extra body fat, while its gut shrinks

Waders

Ardei, Charadrii, Scolopaci

There are plentiful sources of nutrition in shallow sea water and soft shoreline mud, and many different groups of birds have adapted to exploit this rich resource.

Most waders can **swim**, but only avocets have **webbed feet**

Waders have proportionately long legs that allow them to walk through water as they look for food (although many also swim well). This group is defined by shared traits rather than a close evolutionary relationship.

The smaller waders, often known as shorebirds, are related to gulls and terns, and include oystercatchers (*Haematopus*), avocets (*Recurvirostra*), and plovers (Charadriidae), as well as the numerous and variably sized sandpipers (Scolopacidae). Many species breed far inland—the European golden plover (*Pluvialis apricaria*), for example, is a harbinger of spring on the moors of Iceland, where it is said to "sing away the snow." However, most head to sea coasts for the winter where they feed on snails, worms, bivalves, and other invertebrates that live in the nutrient-rich mud of estuaries and bays. Some migrate huge distances to reach good feeding areas. They fly in swirling flocks to safe roosting places when high tide covers the mud. These birds also gather around brackish coastal lagoons, and some species forage on rocky shorelines, tearing mussels and limpets from their moorings and eating carrion and catching flies along the strandline. They are vigilant birds and warn of danger with ringing calls—in the UK the common redshank (*Tringa totanus*) is nicknamed "sentinel of the marshes," while the Eurasian oystercatcher (*Haematopus ostralegus*) is said to call "be wise, be wise" as it takes flight.

Larger waders

The larger waders are a diverse group, including herons and egrets (Ardeidae) and spoonbills and ibises (Threskiornithidae). Some of these species rarely visit the coast, preferring inland waters or drier, grassy environments, but many species of herons and egrets forage at the seashore. They hunt by waiting, very still, until prey appears, but will also chase small fish in the shallows. Spoonbills often join herons and egrets around coastal lagoons and salt marshes. They have long, flattened bills with spoon-shaped tips that they sweep through shallow water, filtering out shrimp and other animals.

CURLEW SANDPIPER
Calidris ferruginea
This elegant sandpiper breeds in the Arctic and migrates to sub-Saharan Africa, stopping at Atlantic coastlines on the way.

BLACK-BELLIED PLOVER
Pluvialis squatarola
This bird has a characteristic "run-pause-peck" hunting style. It catches small fish as well as invertebrates.

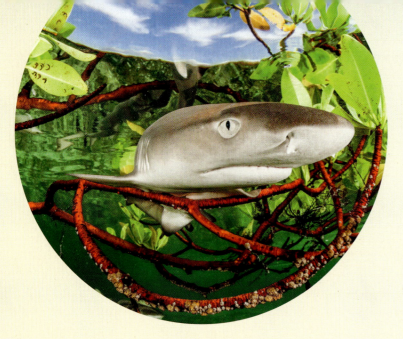

Shark nurseries
Mangrove forests make protective nursery grounds for juvenile animals like this lemon shark (*Negaprion brevirostris*) pup in Eleuthera, Bahamas.

> " You don't cross a mangrove. You'd spike yourself on the roots of the mangrove trees. "
>
> MARYSE CONDÉ, *Crossing the Mangrove*, 1995

Mangroves

Found in tropical and subtropical locations globally, mangroves are habitats consisting of thickets of trees and shrubs growing in salty, tidal, waterlogged conditions often with low oxygen levels.

Few plants can survive in salty, muddy conditions without much oxygen, so hardy mangrove forests provide important coastal habitats. Their root systems offer shelter for marine animals, reduce erosion, and protect the coastline from storms. Tree and shrub species, mainly from the Acanthaceae, Rhizophoraceae, Combretaceae, Lythraceae, and Arecaceae families, are suited to aquatic conditions—their buoyant seeds disperse on the water. Various animal species also thrive and compete in this saltwater environment, including fish, frogs, crabs, sea birds, crocodiles, and even sharks.

Mangroves are under threat from factors including climate change, pollution, and human development. The forests are hugely valuable to local communities: their fish populations provide food, the plants can be used for firewood, timber, and traditional medicine, and some communities use mangrove systems to treat sewage and wastewater.

Mangroves also have spiritual associations. In the Sundarbans—one of the world's largest mangrove forests that stretches from Bangladesh into India—people worship a goddess called Bonbibi, who protects the forest and keeps them safe when they are foraging or collecting firewood.

Above and below
Fish swim around corals that sit under a mangrove forest in Raja Ampat, Indonesia. The country is home to one fifth of the total mangrove area on Earth.

ROOTS THAT BREATHE

The muddy seabed is oxygen-poor, so tree roots need to reach the air to access oxygen. Different root formations have evolved to bridge the gap between the mud and water below, and the air above.

MANGROVE ROOT STRUCTURES AT LOW TIDE

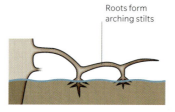
Roots form arching stilts
RHIZOPHORA ROOTS

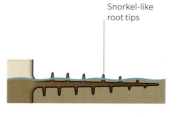
Snorkel-like root tips
AVICENNIA ROOTS

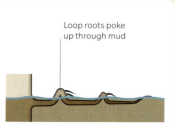
Loop roots poke up through mud
BRUGUIERA ROOTS

Saltwater Crocodiles

Crocodylus porosus

The only crocodile species that regularly swims in the sea, the saltwater crocodile also frequents estuaries, rivers, and swamps. It lives on or close to land around the eastern Indian Ocean, the western Pacific, and Australia, where it is known as the "saltie."

> Saltwater crocodiles can **hold their breath** underwater for **over an hour**

The largest living reptile species, a fully grown male saltwater crocodile can reach 23 ft (7 m) in length and weigh 2,650 lb (1,200 kg); females are smaller. It also has the greatest bite strength of any living animal, thanks to powerful muscles in its head and neck.

Saltwater crocodiles can swim long distances and are sometimes found far out at sea, following ocean currents. More often, however, they swim along shorelines, in and out of estuaries and rivers, or around coastal mangrove forests. They hunt in a wide variety of habitats, so they eat a huge range of prey, from smaller fish, frogs, and crabs, to larger mammals and birds that come to the water's edge. They can launch themselves out of the water to snatch flying birds or bats from the air, or a monkey from a tree branch. The largest adult males have been known to eat kangaroos, buffaloes, and even tigers. The saltwater crocodile is highly territorial and sometimes aggressive. If it attacks humans, it can be deadly—although some of these animals are shy, and avoid humans if they can.

Sacred being

The saltwater crocodile features in many myths and legends. In Timor, crocodiles are sacred and called *abo*, or grandfather; an old story describes how the island of Timor was formed from the body of a seagoing crocodile. In the creation stories of the Iatmul people of Papua New Guinea, the first humans were born from a giant crocodile. Traditionally, this has led to the prows of their large canoes being carved in the shape of a crocodile. In ancient Hindu legends, the ocean god Varuna rides on a crocodile-like creature.

Disguised predator
A saltwater crocodile's greenish eye blends in with the rest of its huge, green-gray, knobbly head, helping it resemble a harmless log when it floats at the water surface.

SALTWATER CROCODILES

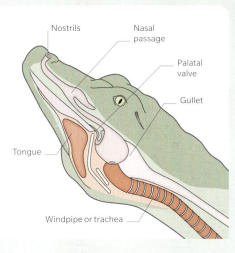

BREATHING SYSTEM

Crocodiles can close off the air supply from their mouths and nostrils separately. While muscles around the nostrils can close them when the crocodile submerges, a valve in the throat—the palatal valve—closes off access to the windpipe from the mouth, allowing the animal to open its mouth and capture prey underwater.

ANATOMY OF A CROCODILE NOSE AND THROAT

Crocodile steed
This 19th-century engraving shows the Hindu god of the oceans, Varuna, riding his "vahana" or vehicle, the crocodile-like Makara.

A female saltwater crocodile can lay up to 90 eggs at a time

Desert Islands

The term "desert island" may suggest the tropical islands of stories and myths, but it actually refers to an island that is uninhabited by humans (in other words, "deserted"). There are around 660,000 islands around the world, of which the vast majority have no human population.

Some of the world's most remote islands originate as chains of submarine volcanoes that grow from the ocean floor and eventually break through the sea surface. They form at the edges of the tectonic plates that comprise Earth's crust.

Some chains, like the Ryukyu Islands off Japan, are "island arcs" that form where one plate slides beneath another. Others, such as the Pacific islands of Hawai'i and the Galápagos, and the Azores in the Atlantic, develop as plates move over "hot spots": points in Earth's mantle (the molten layer underlying the crust) that are hotter than the surrounding areas. The island of Surtsey, which emerged in 1963 in the North Atlantic off Iceland, was formed above the Icelandic hot spot, but also lies in the zone where the North American and

Island nation
One of the world's smallest nations, Palau comprises more than 300 coral islands draped in lush tropical rainforest in the western Pacific.

DESERT ISLANDS 133

Pirate queens
Among the islands of the Caribbean, piracy flourished, reaching its height from the 1650s to the 1730s and giving rise to famous names such as Anne Bonny (seen right in this image, with gun) and Mary Read (left, with sword). In disputes, pirates might abandon one of their number on a deserted island as a punishment.

" 'I remember my poor Mr. Budd always spoke of desert islands as horrid places ... that every one should avoid.' "

JAMES FENIMORE COOPER,
Jack Tier, or The Florida Reef, 1848

Eurasian plates are pulling apart. In October 2023, one of the world's newest volcanic islands emerged in the southwest Pacific, off the coast of Iwo Jima, about 750 miles (1,200 km) from the Japanese mainland. Only 330 ft (100 m) wide, it risks being eroded by waves. Many small islands may be worn away by waves and sink to form submerged guyots, or seamounts—undersea volcanic mountains.

Another form of island develops in tropical or subtropical waters, when a coral reef grows just underwater around a volcanic seamount. In some cases, the volcano dies and is worn away, leaving a ring of coral called an atoll, enclosing a lagoon (see pp.76–77). The atoll may eventually be filled in with accumulations of sediment and organic debris to form an island.

A third form of island forms closer to coasts. These "barrier islands" develop as waves deposit sediment parallel to the shore line, creating single islands or chains that can stretch for hundreds of miles, such as the 113-mile (182-km) Padre Island off the coast of Texas.

Undisturbed habitats
Plants may be carried by air or water to seed themselves on remote islands, including coconuts from coconut palms (*Cocos nucifera*), which are enclosed in a buoyant, water-resistant coating that allows them to float. Birds, bats, and flying insects may carry seeds on their feet or in their droppings. Insects, spiders, turtles, and small reptiles may be carried to the islands on rafts of vegetation. Because of their remoteness, some islands may have no predators and little competition

Reptilian relics
Sea turtles, such as this green sea turtle (*Chelonia mydas*), are a fundamental link in marine ecosystems, helping maintain the health of seagrass beds and coral reefs.

Large flippers can propel the animal at up to 22 mph (35 kph) in water

Non-stop volcano
Te Puia o Whakaari ("the dramatic volcano" in Māori) is one of New Zealand's most active volcanoes. The island is uninhabited due to this activity, but since 1953 it has been a "private scenic reserve."

between species, allowing animals and plants to thrive in isolation. In some cases, this may result in the evolution of unique variants, or species that are endemic (found nowhere else). Alternatively, a species may be preserved on a remote island when it has disappeared elsewhere; one example is the Aldabra giant tortoise (*Aldabrachelys gigantea*), which is endemic to the Seychelles. Desert islands may also be nesting sites or migration stopovers for seabirds; an estimated 15 percent of seabird species nest on coral islands.

Humans and desert islands

In prehistory, people used islands as staging points when crossing oceans. It is known that early humans set out from east Asia across a string of islands in the Indonesian archipelago to reach Australia by around 60,000 years ago. The Polynesian people colonized more than 10,000 islands scattered across 10 million sq miles (26 million sq km) of the Pacific Ocean. Some islands, however, were never settled, or people deserted them, often due to a lack of water or being too distant from other settled lands.

Today, many remote islands are the focus of scientific interest due to their unique habitats and wildlife populations. Some are also under threat from invasive species such as rats, pollution or disturbance by humans, or effects of climate change such as rising sea levels. National and international organizations now recognize the importance of these environments; for example, the United Nations has declared the Phoenix Islands, part of Kiribati in the south Pacific, to be a terrestrial and marine protected area.

The uninhabited Devon Island in northern Canada is the 27th largest island in the world

REMOTE REFUGE

The Juan Fernández islands formed above a volcanic hot spot in the South Pacific. Santa Clara dates from 5.8 million years ago (MYA), Robinson Crusoe from 3.8–4.2 MYA, and Alejandro Selkirk from 1.0–2.4 MYA. The latter two islands are named for the hero of Daniel Defoe's novel, and his real-life counterpart Alexander Selkirk.

THE JUAN FERNÁNDEZ ISLANDS IN A FRENCH ENGRAVING FROM 1842

More than **two-thirds** of the world's **islands** are uninhabited

DESERT ISLANDS   135

Tindhólmur Island
This unpopulated islet is part of the Faroe Islands in the North Atlantic, and is known for its distinctive spiky shape. It features a notable sea arch (see pp.44–47) and sheer basalt cliffs.

Galápagos

A group of volcanic islands renowned for their wildlife, the Galápagos archipelago in the Pacific Ocean straddles the equator 620 miles (1,000 km) from South America.

There are 127 islands making up this remote outpost of Ecuador. They formed above a volcanic hot spot around 4.5 million years ago. While the older islands have now sunk below the sea as the Nazca tectonic plate dragged them away from the hot spot, the youngest islands—Isabela and Fernandina—are still growing. Wolf Volcano on Isabela is the most active of the Galápagos volcanoes, and currently rises 5,610 ft (1,710 m) above sea level. Deep on the adjacent ocean floor, black smoker vents (see pp.294–295) pump hot, metal-rich waters into the ocean.

Marine iguanas
Sun-baked volcanic rocks make an ideal place for these endemic lizards to warm up after feeding in the cold ocean.

Giant tortoises
There are 13 living species of tortoises on the Galápagos Islands, as depicted here in an 1878 engraving. They are thought to have migrated from South America.

The Galápos giant tortoise (*Chelonoidis niger*) is the largest living tortoise species

Europeans first encountered the islands in 1535, when a Spanish ship sailing from Panama to Peru was driven far offshore by strong currents and winds. The name derives from an old Spanish word for saddle, taken from the saddle-shaped giant tortoise shells. By the 17th century, they were a base for pirates, who lay in wait for Spanish galleons. In 1835, Charles Darwin's visit on HMS *Beagle* paved the way for his theory of evolution by natural selection.

Unique flora and fauna

One of at least 2,000 species only found on or around the Galápagos Islands, marine iguanas (*Amblyrhynchus cristatus*) are well adapted to this land of steaming vents, volcanic eruptions, and rocky coastlines. Where competition for land-sourced food is intense, these lizards, with their beaded scales, curved spines, and blunt heads of gnarly cones, have evolved to eat seaweed. Their jaws are lined with three-pronged teeth to spear and rip the seaweed from underwater rocks.

Land iguanas (*Conolophus subcristatus*) are twice the size of their marine counterparts, and giant tortoises are the heaviest reptiles—up to 550 lb (250 kg)—and live the longest. Galápagos racer snakes (*Pseudalsophis biserialis*) are some of the fastest reptiles. Bird life is stunningly varied, with Galápagos penguins (*Spheniscus mendiculus*), lava herons (*Butorides sundevalli*), and Galápagos hawks (*Buteo galapagoensis*) among others.

The islands lie at an ocean crossroads of three main currents, which drive the upwelling of nutrient-rich waters. This attracts plankton, and vast shoals of anchovies, sardines, and black-striped salemas (*Xenocys jessiae*), all of which lead to a feeding frenzy of sharks and whales.

Handwritten labels detail species name and dates

Specimens for study During his voyage on HMS *Beagle*, Charles Darwin amassed a huge number of specimens, including this Galápagos red clingfish (*Arcos poecilophthalmos*).

> " This archipelago is very remarkable: it seems to be a little world within itself. "
>
> CHARLES DARWIN,
> *notes from HMS Beagle*, 1835

CHAPTER 3

On and Above the Surface

Home of winds, waves, and mists, the ocean surface provides habitats for a huge number of life forms, from simple algae to dolphins and sea birds. It has also carried countless voyagers around the globe, from jellyfish riding surface currents to human traders and settlers.

ON AND ABOVE THE SURFACE

Racing on the wind
Two racing yachts under full sail compete in the 1937 America's Cup off Rhode Island. The race takes place every few years between two wind-powered yachts at different locations around the world.

God of the wind
This woodblock print by Japanese artist Utagawa Kuniyoshi, c. 1850, depicts a wind god, representing the trigram symbol *Xun* (Wind) in Chinese divination.

Wings have a lock at the shoulder, enabling the bird to glide on wind currents without flapping

Bill has tubes running along each side, to help the bird register air speed during flight

Ocean Winds

Winds have existed since the ocean and land first formed more than four billion years ago. They control the daily process of weather, buffer the planet from extremes of climate, and drive ocean currents that redistribute heat from equator to poles.

Wind occurs when bodies of air move from areas of high pressure to areas of lower pressure in the atmosphere. Such pressure differences arise from the way in which the sun's radiation heats the Earth. Due to the planet's spheroidal shape, heat is concentrated at the equator, where the Sun lies directly overhead, whereas at higher latitudes—farther north and south—it diffuses over a wider area and over a thicker layer of atmosphere. So, at the poles, the atmosphere loses more heat than it gains from the sun, especially in winter, which makes these areas the coldest on Earth. These differences in temperature cause air to move: as air is warmed it becomes less dense and rises, leaving an area of low pressure beneath, into which denser, cold air flows.

In addition, Earth's rotation and the Coriolis effect (see pp.22–23) cause the air to be deflected laterally as it moves, giving easterly or westerly winds (named for the direction they come from). These movements of air across Earth form horizontal bands known as cells. The northern

Riding the wind
Albatrosses (see pp.144–145) use wind to travel large distances. An albatross may cover 10,000 miles (16,000 km) and remain airborne for more than a year in a single flight.

Offshore wind farm
Wind is an inexhaustible source of renewable energy. Offshore wind farms, such as this one off the coast of the Netherlands, can generate more electricity than onshore turbines, due to the higher wind speeds over the oceans.

and southern hemisphere each have three of these cells. The largest ones, extending from the equator to about 30° latitude, are known as the Hadley cells; the smallest ones, over the poles, are called polar cells; and situated between the two, from 30° to 60° latitude, are the Ferrel cells.

In the Hadley cells, warm air at the equator rises and flows toward the poles, cooling and sinking at the cell boundary. In the northern hemisphere, the Coriolis effect causes high-level air to move eastward; at lower levels, there is a corresponding air movement from east to west, giving the easterly winds known as trade winds. Warm air carries a lot of moisture; as it rises,

clouds form and produce rain and storms. The trade winds meet at the equator in a low-pressure area called the Intertropical Convergence Zone (ITCZ), historically known as the doldrums.

The polar cells lie from the poles to 60° latitude: here cold, dense air at the poles sinks and flows toward 60° latitude, where it warms and rises. In the upper levels, air moves east due to the Coriolis effect, while at low levels the air moves from east to west, giving the polar easterly winds.

Between these cells are the Ferrel cells, in which some of the air rising at 60° latitude flows toward the equator; at 30° latitude it meets the air from the Hadley cells, and sinks. The air movement in the Ferrel cells is influenced by contact with the other two cells rather than direct heating. The surface winds at these mid-latitudes are westerly, often carrying abundant clouds and rain.

Harnessing ocean winds

Humans made their first sea voyages more than 50,000 years ago, sailing rafts from southeast Asia, via islands in the southwest Pacific, to New Guinea and Australia. The first ships with sails appeared in Egypt around 3500 BCE. Travel by sea became vital for people migrating and forging trade routes.

Wind and storms made seafaring dangerous, but the wind also provided power for fishermen and sailors, so people feared and revered it. In many cultures, the winds were seen as deities. Notably, the word "hurricane," for a tropical cyclone (storm) in the Atlantic, comes from Huracán, the Mayan god of winds, storms, and floods, who helped in the creation of humanity.

Today, ocean winds provide power for energy as well as transportation. There are more than 290 offshore wind farms, producing a growing proportion of the world's renewable energy.

Whirling waterspout
This engraving from *Barnes's Complete Geography*, by James Monteith was published in 1886. A waterspout is a fast-rotating column of air that forms a funnel-shaped cloud above the water.

> " ... When we have laugh'd to see the sails conceive
> And grow big-bellied with the wanton wind ... "
>
> WILLIAM SHAKESPEARE,
> *A Midsummer Night's Dream*, c. 1594

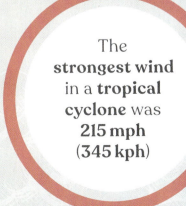

The **strongest wind** in a **tropical cyclone** was **215 mph (345 kph)**

COASTAL WINDS

Smaller-scale air currents are generated at the shorelines. The land heats up faster than the ocean during the day. As warm air rises, it draws in cool air from the sea; this creates an onshore wind. At night, the land cools down more quickly, reversing the flow of air, and an offshore wind occurs.

AIR MOVEMENT

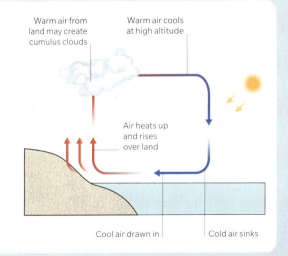

Albatrosses

Diomedeidae

Few birds can rival albatrosses for their effortless mastery of the marine environment. Long-lived and far-ranging, they are exceptional avian explorers.

Albatross assembly
Black-browed albatrosses (*Thalassarche melanophrys*) are often solitary, but large groups may form around concentrations of food.

Albatrosses, along with around 100 other bird species, are part of the Procellariiformes order. These birds are true seafarers, never more at ease than when skimming or riding the waves of the ocean, or soaring on gusts of wind. Albatrosses have many unusual characteristics: they are among the slowest birds to reach adulthood, the longest-winged, and the longest-lived. They breed mainly on tiny islets at various locations in the southern hemisphere, with pairs reuniting after long months apart with intense displays of strutting, bill-clacking, and posturing before the annual process of producing and incubating one egg. Over the following months, both parents tend to the growing chick. Larger species may only breed once every two years. Since their breeding sites have

A **snowy albatross's wingspan** can reach 12 ft (3.7 m)— the **biggest** of any bird

no native predators, the parent birds do not attempt to hide the nest or chick, and typically show no fear of human visitors.

Their demeanor on land, coupled with the ungainly gait of a bird built to fly, not walk, inspired nicknames such as "gooney" and "mollymawk"—Dutch for "foolish gull"—and also meant albatrosses were exploited for food by early explorers. However, they were welcomed by seafarers when encountered on long voyages—an albatross keeping company with a ship meant good luck, and sailors would throw scraps to encourage them to linger. *Rime of the Ancient Mariner*, a poem by Samuel Taylor Coleridge published in 1798, describes a sailor who shot an albatross and was forced to wear it around his neck as punishment. The saying "albatross around your neck" describes a mistake or burden that continues to haunt and impede one's life, long after the event.

Effortless endurance

Flying over sea is different to flying over land. Thermals (rising warm air currents) form over hilly landscapes, allowing broad-winged birds such as eagles to soar effortlessly. At sea, air currents form between wave crests, and long-winged albatrosses are adept at steering through these updrafts to capture lift, allowing them to remain airborne without a single wingbeat. The detail of this dynamic soaring was revealed by computer modeling in 2017. It allows albatrosses to cover huge distances in the air without rest, but they can sit on the sea surface at any time.

Venerable but vulnerable

As albatrosses mostly feed from the sea's surface, they are vulnerable to ingesting plastics and to taking bait from longlines—birds can drown as the lines are pulled under and they get trapped, unable to resurface. Introduction of nonnative predators to their breeding islands also spelled disaster. There are many conservation efforts in operation to support albatross populations. Ornithologists working with Laysan albatrosses on Midway Atoll on the Hawaiian archipelago look forward each year to the return of one particular breeding female. At the time of writing, she is at least 74 years old. "Wisdom," as she is known, has had her leg ring replaced six times since it was first fitted, and over her lifetime has raised more than 30 chicks. She is also likely to have flown at least 3 million miles (4.8 million km).

WAVED ALBATROSS
Diomedea irrorata
Nearly 35,000 pairs of this large, strikingly long-billed albatross breed on Española Island, in the Galápagos archipelago.

BLACK-FOOTED ALBATROSS
Phoebastria nigripes
Sometimes one black-footed albatross incubates for more than a month, before its partner returns to take over.

The dark "frown" helps the albatross cope with bright glare from reflected sunlight

Buller's albatross (*Thalassarche bulleri*)
This species breeds only on islands around New Zealand, although it flies across to South America outside of the breeding season. The largest colonies are found on the Chatham archipelago, off the eastern coast of South Island.

Eye of the storm
This view of Typhoon Trami, a tropical cyclone in the west Pacific, was taken from the International Space Station in September 2018. The spiraling movement of clouds around the eye is clearly visible.

Tropical Cyclones

Known as cyclones, typhoons, or hurricanes, the planet's biggest oceanic storms go by many names, but all share a similar power, destructive capacity, and deadly unpredictability. These extreme weather events are increasing in size and frequency.

The largest, most violent storms have different names around the world: cyclones, meaning "coiled serpents," in the Indian Ocean and South Pacific; typhoons after the Chinese *tai fung*, or great wind, in the northwest Pacific; and hurricanes in the Atlantic and northeast Pacific, after the Central American storm god, Huracán. However, behind the myths, these are all tropical cyclones. They are created by the overheating of the oceans, specifically when water temperatures exceed 80°F (27°C) in the upper 150 ft (45 m) of the sea.

As water evaporates, moist air rises and then condenses as it cools to form billowing cumulonimbus clouds. Air rushes inward across the sea surface, evaporating more water as it whirls past, and begins to spiral upward around a calm eye of the storm.

Wind speeds near the wall of the eye can exceed 220 mph (350 kph). As the storm grows, it becomes self-sustaining, sapping heat energy from the ocean and driving the spinning winds ever faster. Spiraling bands of thunderstorms

Two serpents frame face

Tláloc bowl
This 14th–16th-century stone bowl from Mexico depicts Tláloc, the Aztec god of rain and thunder, who can provoke hurricanes but also brings earthly fertility, providing life and sustenance.

TROPICAL CYCLONES   147

> On average, there are **80 tropical cyclones** a year. In very **warm years** that can double

INSIDE A HURRICANE

An increase in sea surface temperature causes vast amounts of water vapor to evaporate, creating huge convection towers of billowing cumulonimbus clouds up to 10 miles (16 km) high. As the hot air rises, more air is dragged across the surface to fill the void. The Coriolis effect created by Earth's rotation causes the winds to spin and the clouds spiral upward.

CROSS SECTION OF A HURRICANE

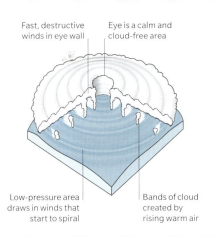

Fast, destructive winds in eye wall
Eye is a calm and cloud-free area
Low-pressure area draws in winds that start to spiral
Bands of cloud created by rising warm air

The storm god
This c. 1880 woodcut by Yoshitoshi Taiso shows Susanoo-no-Mikoto, the Shinto god of the sea and storms. A powerful figure, he is often portrayed as temperamental, sowing chaos.

begin to rotate around the center, counterclockwise in the northern hemisphere and clockwise in the south due to the Coriolis effect (see p.23). The eye can measure 15–25 miles (25–40 km) across. The most destructive winds are confined to a radius of 60 miles (100 km), while the whole complex can be over 900 miles (1,500 km) wide.

The progress of the storm is slow and erratic, making it hard to predict where it will strike land. As it tracks inland it loses power, but its potential for destruction lingers on.

Power and change

The energy released in a single day by wind, thunderstorms, and lightning is enough to power the industrial production of the US for a year. The damage it wreaks is extreme and feared by vulnerable coastal communities, as well as by those at sea. While people once invoked gods as the masters of these forces of nature, humans are playing a growing role. Hotter seas caused by human-made climate change are making tropical storms more frequent and powerful around the world. In 2024, warm water temperatures in the Atlantic combined with shifting weather patterns led to a particularly intense hurricane season in the second half of the year, with 11 hurricanes recorded—four more than average.

Waves

Waves are movements of energy rippling across the ocean. Most are driven by the wind disturbing the sea's surface, and the gravitational pull of the sun and moon, producing tides, can also cause waves.

Wind-generated waves arise from the friction of wind blowing across the surface of the water. The disturbance causes water molecules at the surface to roll back and forth, up and down, and slightly forward in a circular motion, producing ripples. The greater the wind, the bigger the wave, and the more energy it releases when it breaks.

Ghost warriors
This woodblock triptych (c. 1849–1852) by Utagawa Kuniyoshi depicts a sea battle between the Taira and Minamoto clans, in which the ghosts of the drowned Taira rise up to attack Minamoto general Yoshitsune's ship during a storm.

to larger, irregular waves, is known simply as "sea." As these waves leave the region in which they were generated, the longer, faster waves outpace the shorter ones, and gradually fall in with other waves traveling at similar speed. Eventually, a regular pattern, or "swell," develops that remains constant as it travels out across the ocean for hundreds or thousands of miles.

When waves meet shallower water at the shoreline, the energy starts to affect the water closest to the seabed, and the circular movement is slowed and distorted by friction. This causes the waves to form peaks, which collapse on the shore (see p.150).

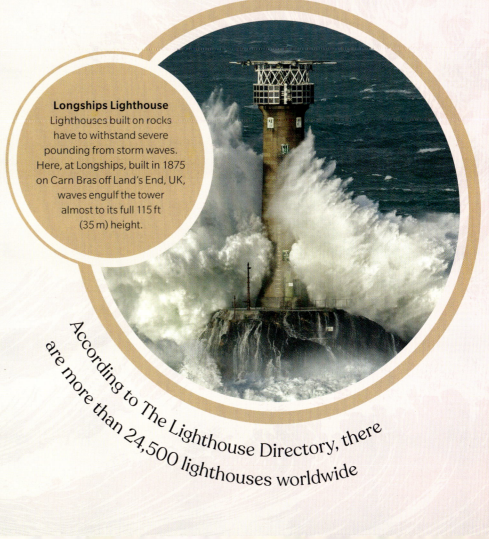

Longships Lighthouse
Lighthouses built on rocks have to withstand severe pounding from storm waves. Here, at Longships, built in 1875 on Carn Bras off Land's End, UK, waves engulf the tower almost to its full 115 ft (35 m) height.

Wind speed, duration, and the distance over which the wind blows, known as the "fetch," all influence the growth of waves. The smallest waves, formed by gentle puffs of wind, are tiny, diamond-shaped ripples known as cat's paws. Only some waves build into maturity. These waves go through three stages: sea, swell, and surf. The state of random choppiness, building

According to The Lighthouse Directory, there are more than 24,500 lighthouses worldwide

Power of nine
Depicting humanity's struggle against nature, *The Ninth Wave* (1850) by Ivan Aivazovsky takes its name from a term for a huge wave that follows a series of increasingly bigger waves.

Major storm waves generated in the Antarctic Ocean can take nearly a week to cross the Pacific before they break as surf along the shores of Hawai'i, or travel for another few days before quietly washing ashore on the beaches of Alaska. The original pattern of swell that first left the Antarctic Ocean remains intact.

Some areas of the ocean are notorious for huge waves. In the Bay of Biscay, off western France and northern Spain, areas of low air pressure (depressions) from the open Atlantic give rise to powerful wind storms. The Grand Banks of Newfoundland comprise shallow waters, just 167–328 ft (51–100 m) deep, in an area often subjected to tropical storms. Most dangerous of all is the Antarctic Ocean, stirred by fierce westerlies and swept by the powerful Antarctic Circumpolar Current.

A storm surge is a rise in the sea level above the expected tidal height. Strong, fast storm winds transfer increased amounts of energy to the sea surface; if the fetch (the distance over which the wind blows) is long, or the storm lasts for a long time, this can create huge storm waves.

There are also gigantic individual waves known as "rogue waves," which strike erratically and disappear without a trace. When two or more storm waves merge or encounter an opposing fast-flowing current, such as the powerful Agulhas Current off eastern South Africa or the Kuroshio Current off the coast of

Spiral shell shape forms the body of the bottle

Scent of the sea
The stopper on this art nouveau–style perfume bottle represents Amphitrite, the wife of Poseidon, Greek god of the sea. A Nereid, or sea nymph, she had the power to calm the waves.

> " Waves are the practice of the water ...
> Water and waves are one. "
>
> SHUNRYU SUZUKI, *Zen Mind, Beginner's Mind (50th edition)*, 1970

WHEN SEA MEETS SHORE
As a series of waves enters shallow water, the circular motion of the water molecules is disturbed by friction with the seafloor. The movement of the water becomes elliptical, and the top of the wave starts to move faster than the base. This causes the front of the wave to become steeper than the back, and this unstable structure finally topples forward as surf.

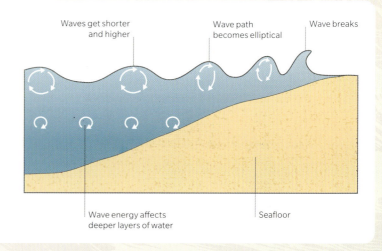

WAVES BREAKING ON THE SHORE

Japan, rogue waves can suddenly tower to four times their original height and break at sea under their own instability. Even large ships can capsize when facing such power.

Waves in myth and legend

For thousands of years, humans have admired and feared the power of the waves. They have seen waves as instruments of the gods, or even as deities themselves. Some of the folklore passed down through the centuries has been found to have its roots in real events.

In Norse mythology, the sea is ruled by the jötunn (giant) Ægir and the sea-goddess Rán, and their nine daughters personify different aspects of the waves, with names such as Bára (billow) and Himinglæva (shining like heaven).

According to Irish myth, waves come in groups of nine. The ninth wave—the last and most powerful—is the boundary between the human world and the Otherworld, and is said to carry away the souls of the dead.

In the *Mahābhārata*, the Sanskrit epic revered in the Hindu faith, the god Krishna lived in the beautiful city of Dvārakā, and when Krishna departed to the spirit world, the sea rose and the city was consumed by the waves. The story was considered to be legend, but during the 1980s India's National Institute of Oceanography carried out excavations near the coastal city of Dwarka in modern-day Gujarat, and found the remains of an underwater fortified city dating from at least 1500 BCE.

> In the winter of 2013–2014, **storm waves** off the west coast of Ireland lifted a **boulder** weighing **683 tons (620 metric tons)**

Hawaiian surfers, 1960
Surfing originated among the Polynesian peoples, but from the 1960s it became popular with Westerners in places with huge waves, such as Hawai'i.

In the jaws of the ocean
A surfer rides a huge wave at Praia do Norte in Nazaré, Portugal. The beach is home to some of the largest waves in the world due to the effects of the underwater Nazaré Canyon, which starts just off the shore and extends for 140 miles (230 km) out into the Atlantic, reaching depths of up to 16,000 ft (5,000 m). The canyon channels and amplifies ocean swells, creating waves up to 100 ft (30 m) high from October to March each year.

Tsunamis

A tsunami is a large, fast-moving wave, or series of waves, caused by a sudden disturbance in the water such as an undersea earthquake. When a tsunami reaches the shore, it can cause large-scale destruction.

> The 2004 **Indian Ocean** tsunami created waves as high as **100 ft (30 m)**

Tsunamis are not like normal ocean waves, which are caused by the wind over long distances and only travel on the surface of the ocean. Instead, an earthquake on the seabed, or another disturbance such as a huge landslide or volcanic eruption, suddenly displaces a large volume of water. This causes shock waves that spread out in all directions and travel though the entire water column, from the ocean floor to the surface. On the surface, the water movement sets up large, ripple-like tsunami waves, which spread out at high speed, until they dissipate or reach land. When the latter happens, most tsunamis do not resemble a single, giant breaking wave. Instead, they usually comprise a series of waves that surge ashore quickly, or a sudden, dramatic rise in sea level that swamps the coast. The powerful rush of water can sweep away people, trees, and vehicles; flatten buildings; and carry boats and debris far inland. It is extremely dangerous to be swept along in tsunami water, because the debris it

TSUNAMIS

SPEED AND MAGNITUDE
Once triggered by seismic activity, the tsunami's waves move quickly through deep water at up to 500 mph (800 kph). On the surface, they may only be around 3 ft (1 m) in height, and are often hundreds of miles wide. Once the waves reach shallow water, they slow in speed and increase in height.

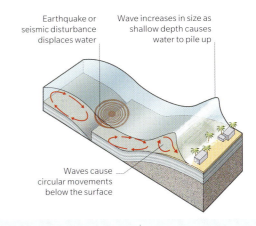

The Meiji Sanriku tsunami
This woodblock print by Utagawa Kokunimasa, depicts the effects of a tsunami that struck northern Japan in June 1896 and killed around 22,000 people. It shows how a woman was carried along in her bathtub, ending up on top of a hill.

carries can cause injury or trap people underwater. Afterward, when the wave retreats, it can also sweep people back out to sea with it.

Human toll
Tsunamis have been responsible for some of the worst natural disasters. The deadliest on record was the 2004 Indian Ocean tsunami, triggered by an undersea earthquake near Banda Aceh in Indonesia. Around 230,000 people died in the quake and the resulting tsunami. However, the Moken people, who live on islands in Thailand and Myanmar, recognized the warning signs of a tsunami from the legend of Laboon—a huge wave sent by angry ancestors—and moved to higher ground before the first wave hit land.

> " I went to see one of those pianos drowned in tsunami water ... it was totally out of tune ... I thought, 'Nature tuned it.' "
>
> RYUICHI SAKAMOTO, Japanese pianist and composer

Sea wall overspill
Rising seawater surges over a sea wall in Miyako, Japan after a huge earthquake caused a powerful tsunami there in 2011.

LAUGHING GULL
Leucophaeus atricilla

This widespread, dark-backed gull breeds mainly on Atlantic coasts of North America, after which some migrate south of the equator. In winter, its black hood mostly disappears.

SILVER GULL
Chroicocephalus novaehollandiae

Common on coasts all around Australia, the silver gull is an opportunistic species that often forages in urban areas. It mainly breeds on undisturbed offshore islands.

IVORY GULL
Pagophila eburnea

Found in the high Arctic, and often seen resting on icebergs, this small, dainty gull's white plumage is often stained by whale carrion and the remains of other dead animals.

Adaptable feeders
The kelp gull (*Larus dominicanus*), a large species found along southern hemisphere coastlines, has adaptable feeding habits. It will even peck blubber from living whales.

Long, relatively narrow wings allow gulls to glide with ease on air currents that form between wave crests

Dark eyes can be seen on most larger species (genus *Larus*). These become paler and brighter as the bird approaches maturity

Leg color can be helpful for gull identification, although the legs are less brightly colored outside of the breeding season

" He was not bone and feather but a perfect idea of freedom and flight. "

RICHARD BACH, *Jonathan Livingston Seagull*, 1970

Gulls

Laridae

The most adaptable seabirds, gulls are found along coastlines and at sea all around the world. Their ability to thrive alongside humankind makes them a familiar sight.

> Utah's state bird is the **California gull.** Legend tells of them **saving a harvest**

There are about 50 species of gull, ranging in length from the insect-eating little gull (*Hydrocoloeus minutus*) at 11 in (29 cm), to the great black-backed gull (*Larus marinus*), a formidable predator at 31 in (79 cm). Some, such as the kittiwakes (*Rissa*), live mostly offshore, while others occur mainly inland—the term "seagull" is often a misnomer. Most, however, spend at least the nonbreeding season feeding from seashores and inshore waters on fish and other sea life, as well as land animals, carrion, and edible items discarded by people.

Wise and watchful

Gulls are gregarious, nesting in colonies and foraging in flocks, but they also form strong and lasting pair bonds. Being conspicuous when nesting, they are loud and ferocious in defense of their eggs and chicks, often dive-bombing intruders to scare them off. Larger gulls also raid the nests of smaller species.

Gulls are intelligent birds, as indicated by their opportunistic and social nature. Although they prefer natural food, in coastal towns they regularly steal food from humans. Studies show that they prefer food that humans have handled. Many raids involve a swoop from behind, but making eye contact discourages them. Gulls quickly learn to recognize individual humans. The most specialized species is the swallow-tailed gull (*Creagrus furcatus*) of the Galápagos and nearby islands. This nocturnal gull hunts for fish and squid as they come to the surface at night. Another Galápagos species is the rarest—the lava gull (*Leucophaeus fuliginosus*). It has an unusual dark, smoky plumage and is more solitary than other species. Gulls' closest cousins are terns (see pp.166–167), skimmers, and skuas, which all tend to hunt on the wing.

Superstition

Many coastal communities tell stories of gulls carrying the spirits of lost seafarers, and sailing tradition dictates that gulls must not be harmed. The 6th century Welsh saint Cenydd (Kenneth) was said to have been saved by black-headed gulls as a baby, after he was cast adrift in a basket. His feast day is marked by displaying a gull effigy on the tower of the church he founded in the village of Llangennith.

Following the boats
Gulls of various species flock around fishing boats for the chance to eat discarded bycatch. They squabble fiercely over choice morsels, with larger species usually dominating.

Sargasso Sea

The Sargasso Sea in the North Atlantic is the only sea in the world without a land border. Instead, it is surrounded by ocean currents, and its boundaries move as ocean streams change.

> The **Sargasso** Sea is home to the endangered **Bermuda petrel**

This 2-million-sq-mile (5.2-million-sq-km) body of water is located off the southeastern seaboard of the US and northeast of the Bahamas; the sea also encompasses Bermuda. Lying within the Northern Atlantic Subtropical Gyre, the Sargasso Sea is bounded by the Gulf Stream to the west, the North Atlantic Current to the north, the Canary Current to the east, and the North Atlantic Equatorial Current to the south. Within the gyre, the water is calm and crystal clear. Portuguese sailors in the 15th century first named the golden *Sargassum* seaweed, with its

Shelter and food
Filefish (Monacanthidae) use the mats of *Sargassum* for shelter, and also feed on invertebrates living in the seaweed.

Turtle hatchlings
Some research suggests that loggerhead sea turtles (*Caretta caretta*) and other baby sea turtles spend their "lost years"—after they hatch and before they return to shore—in the Sargasso Sea.

clusters of air-filled bladders, using their word for bunches of small grapes. Unlike other seaweeds, *Sargassum* spends its entire life floating freely in thick rafts. Oceanographer Sylvia Earle described the Sargasso Sea as a "golden floating rainforest."

ships would become entangled in the seaweed forever. Even into the 20th century, legends persisted that ships and aircraft would disappear in the "Bermuda Triangle." However, it is the sea that is under threat from human activity in the form of pollution from waste and noise from shipping. To protect it, in 2014 an international declaration was signed by nations agreeing to protect and conserve this unique marine environment.

Humpback whales (*Megaptera novaeangliae*) migrate through the Sargasso Sea each year

It is home to around 145 invertebrate species and 127 species of fish, as well as the Sargassum swimming crab (*Portunus sayi*) and Sargassum shrimp (*Latreutes fucorum*). Green sea turtle (*Chelonia mydas*) hatchlings feed and shelter in the weed, while it also provides a hiding place for predators such as the sargassum angler fish (*Histrio histrio*). Marlin (Istiophoridae), dolphinfish (*Coryphaena hippurus*), porbeagles (*Lamna nasus*), and great white sharks (*Carcharodon carcharias*) use the area to spawn or breed. The European eel (*Anguilla anguilla*) migrates up to 6,200 miles (10,000 km) across the Atlantic to the Sargasso Sea to spawn.

Sea of mystery
For centuries, sailors viewed the Sargasso Sea as a place of mystery or even danger. Christopher Columbus recorded his sailors' fears that their

Perfect camouflage
Pictured in Frederick Nodder's 1796 illustration, the sargassum angler fish—a type of frogfish—camouflages itself in the seaweed and can change color to match its background.

Specialized fins allow the fish to "walk" along the seaweed and to jump out of the water onto clumps of *Sargassum*

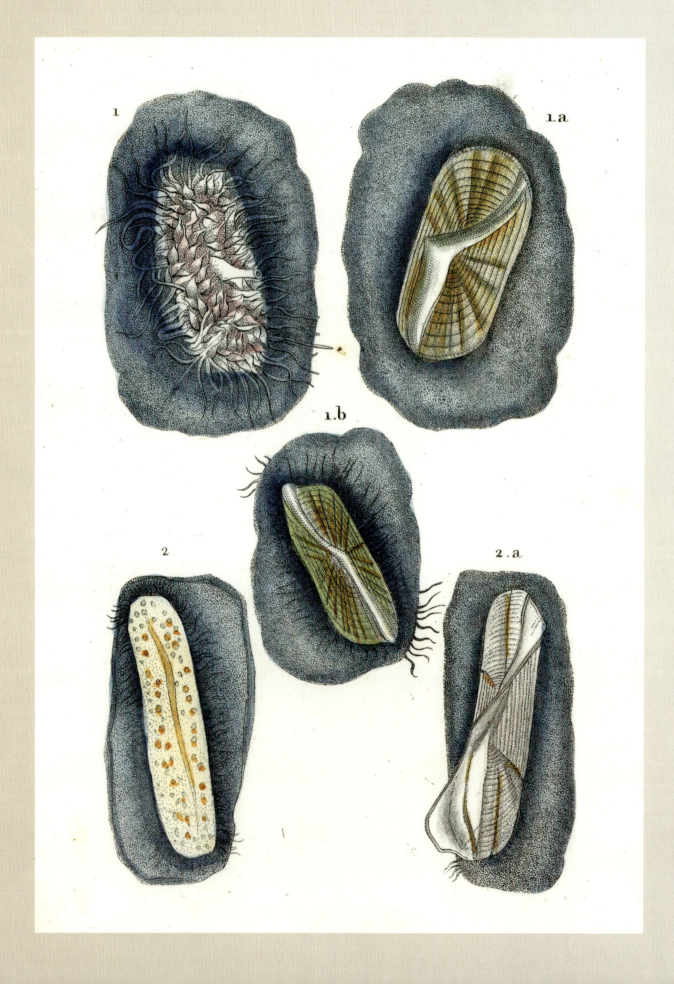

Early observation
This hand-colored engraving was drawn by natural history artist Jean-Gabriel Prêtre (1768–1849). Prêtre was well known for his illustrations of the natural world, and had some animal species named after him.

The sail position affects the direction in which a by-the-wind sailor will drift

By-the-Wind Sailor

Velella velella

This unusual species is related to jellyfish and corals. It floats in the ocean at the mercy of the winds, and often washes up on shore.

Looking blue
By-the-wind sailors' distinctive blue pigmentation helps protect them from the glare of the sun's ultraviolet rays. It also provides camouflage, helping them blend in with the ocean.

Like the Portuguese man-of-war (*Physalia physalis*) and the blue button jellyfish (*Porpita porpita*), the by-the-wind sailor is a free-floating hydrozoan. Its other common names include sea raft, purple sail, or little sail, and it is also known as *Velella*. It has a blue, disk-shaped body of up to around 4 in (10 cm) that floats on the water and a "sail" that sticks out on top.

Life at sea

By-the-wind sailors are in the same group as corals and anemones: Cnidaria. Although this jellyfish-like creature looks like one animal, it is made up of a colony of tiny polyps that work together to catch food. There are different types of polyp and each one has a different role. Gastrozooids are responsible for feeding, gonozooids for reproduction, and dactylozooids for stinging prey and self-defense. The tentacles do not offer protection from its main predators, blue dragon sea slugs (*Glaucus atlanticus*) and purple snails (*Janthina janthina*), which are unaffected by its stings.

By-the-wind sailors float across the surface of warm to warm-temperate waters where plankton grow. As well as capturing plankton in the stinging tentacles, by-the-wind sailors have an extra food source. Like corals, they have a symbiotic relationship with microalgae called zooxanthellae. These tiny algae live inside the organism, producing energy by photosynthesis, and sharing this food with their host.

A NOMADIC PREDATOR

The body of a by-the-wind sailor is an oval disk that floats on the water with air-filled bladders that make it buoyant. Its sail catches the wind, propelling it across the ocean. Its stinging tentacles are suspended in the water, ready to catch any plankton.

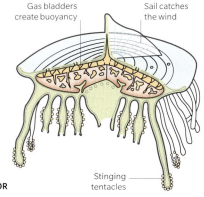

CROSS SECTION OF A BY-THE-WIND SAILOR

Whirlpools

Whirlpools are rotating bodies of water caused by opposing currents. These naturally occurring marine phenomena have long been the inspiration for seafarers' stories and mythical tales.

Whirlpool power
An aerial view of whirlpools forming as tidal currents roar through Saltstraumen in Norway. Most vessels avoid the channel, but sightseeing boats take tourists to view the whirlpools.

In Homer's *Odyssey*, Odysseus and his men have to sail between the many-headed monster Scylla, and the mighty Charybdis, a mythical whirlpool described as so deep that the dark seabed can be glimpsed at the bottom. Odysseus relates how he survived the current by grabbing a branch of a tree growing from a rock above the sea as his raft was sucked in.

The inspiration for this mythical peril is an area of sea located between Sicily and mainland Italy. In reality, whirlpools are not as powerful or deadly as Charybdis—but, in common with it, they are found in narrow seaways and straits. As the tide flows in and out, the mass of seawater rushing through a narrow channel can create a powerful current. Sometimes, because of two opposing currents interacting, or because of obstacles or the shape of the seabed, the current spirals into a rotating area of water—a whirlpool, also known as a maelstrom. Additionally, a vortex is the technical name for a whirlpool that has a downcurrent (or, in the air, can be used in relation to tornadoes). The size of the whirlpool is partly determined by the speed of the currents that cause it.

> " Not very far from the shore ... is that very deep whirlpool of waters which we call by its familiar name 'the navel of the sea.' "
>
> PAUL THE DEACON, *History of the Lombards*, c. 790

Tsunami aftermath
A massive tsunami that struck Japan's east coast in March 2011 created this giant whirlpool, as the enormous surge of seawater flowed quickly ashore, then back again.

Whirlpools of the world
Some of the world's most powerful whirlpools are found at Saltstraumen in Norway, where the current can flow at up to 25 mph (40 kph). The whirlpools there can be 33 ft (10 m) wide, and strong enough to suck in a swimmer or overturn a small boat—but not strong enough to swallow a large ship. Other major whirlpool locations include the Naruto Strait in southern Japan; the Old Sow, between Deer Island in New Brunswick, Canada, and Moose Island in Maine, US; and Te Aumiti (French Pass) at the north tip of New Zealand's South Island.

Mighty maelstrom
This c. 1853 woodblock print by Utagawa Hiroshige is of Japan's Naruto whirlpool. As is often the case, the maelstrom is surrounded by violent waves as a result of the speeding currents.

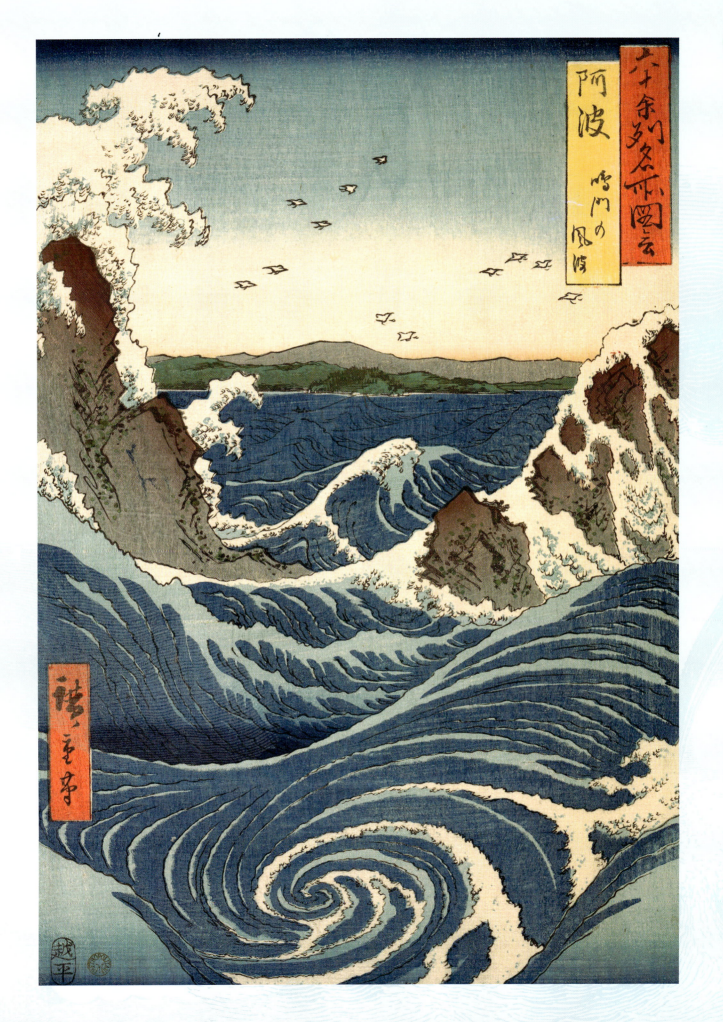

164 ON AND ABOVE THE SURFACE

Sea Mists and Fogs

A common weather phenomenon at sea, these low-level clouds are a well-known danger to seafarers, obscuring hazards such as cliffs and rocks. However, they have also inspired stories, myths, and legends.

Impression, Sunrise
Claude Monet's painting of a morning sea mist over Le Havre port, France, shocked art critics when exhibited in 1874, and helped establish the Impressionist art movement.

SEA MISTS AND FOGS

San Francisco fog
Sea fog lingers beneath the 745-ft- (227-m-) high towers of the Golden Gate suspension bridge, which spans the strait connecting San Francisco Bay and the Pacific Ocean.

> "The sea fog originates in the cooling of air by contact with the colder surface of water..."
>
> Annual Report of the Board of Regents of the Smithsonian Institution, **1896**

Sea mist and fog are both types of low-lying clouds that form when warm, moist air moves over a colder sea surface. The water vapor held in the air cools and condenses around minute particles, forming tiny water droplets that remain suspended as clouds. Different types of particle can seed clouds, from dust blown from the land, to ice in cold weather, and salt crystals in the ocean spray. When brown kelp (Phaeophyceae, a type of seaweed) becomes stressed, it can release iodine particles, which can also act as cloud seed.

The only difference between mist and fog is the density of the suspended water droplets and hence the visibility. If visibility is more than 3,300 ft (1,000 m), the clouds are mist; if visibility is less than that, they are fog. Both types are common along the coastline, especially in spring and summer when the land heats up quickly and warm air blows across the cooler water. The mist or fog lingers until the heat of the Sun burns it off or a crisp wind blows it away.

The impact of mist and fog on coastal communities has spawned a number of words and stories: a cold sea fog is known as a fret in England and a haar in Scotland. In Norse mythology, Niflheim is the world of mists, and Valkyries are female figures who ride the sea mists, one of whom is named Mist, derived from the Old Norse word *mistr*. Celtic and Irish folktales portray mist as a veil of invisibility.

Dangerous conditions

Wherever cold coastal currents occur, sea fog is common. Grand Banks off Newfoundland, Canada is known as the foggiest place in the world, experiencing more than 200 days of fog per year. This is caused by the cold Labrador Current meeting the warm Gulf Stream. Icebergs in the area make it a particularly hazardous part of the ocean to navigate.

By contrast, it is the combination of thick sea fog, strong winds, heavy swells, and shifting sands that make the waters along the Skeleton Coast (see pp.104–105) in Namibia so dangerous for shipping. The cold Benguela Current that powers northward along the coast meets warm air spilling off the Namibian Desert, giving rise to dense and persistent sea fogs that can extend up to 62 miles (100 km) inland.

Such "fog deserts" are known all around the Arabian peninsula, and also in Chile, where the glacial waters of the offshore Humboldt Current spawn dense sea fogs that creep inland across the Atacama Desert. This in turn can create fog oases where plant life is able to thrive because of the water the fog supplies.

Donner swings his hammer to clear the mist

Divine dispersal
God of thunder Donner (Thor) commands the swirling mists to disperse in Arthur Rackham's 1910 illustration for *The Rhinegold and the Valkyrie.*

166 ON AND ABOVE THE SURFACE

Long, pointed wings give terns fast but agile flight, and their bodies are lightweight enough for them to sustain flapping flight for long spells

Forked tails, often with long outer "streamers," create a characteristic outline that helps distinguish terns from gulls

Because of their **long, forked tails**, **terns** are nicknamed **"sea swallows"**

Life in the air
Even when not undertaking their epic pole-to-pole migrations, Arctic terns tackle most tasks in mid-air. This includes noisily settling disputes with neighbors at breeding colonies.

Double act
John James Audubon, author of the seminal *Birds of America*, was first to describe and illustrate the snowy-crowned tern (left) in 1838. It is shown here with a Forster's tern, its closest relative.

Terns

Sternidae

Terns are graceful seabirds whose flight appears almost effortless. They are close cousins to the gulls in evolutionary terms, but differ from them in habits and habitats, as well as in appearance.

Terns live in large colonies of bustling activity and noise. Potential predators are warned off by dive-bombing attacks from dozens of birds—terns are small, and their colonies conspicuous, so they ward off danger through ferocity. Their aerial skill and long, sharp bill are usually employed to catch fish or pick prey from the water's surface. Terns live life on the wing. Many species undertake epic migrations between breeding and wintering grounds. The most northerly-breeding Arctic terns (*Sterna paradisaea*) may fly more than 60,000 miles (97,000 km) in a year on their journey from the Arctic to the Antarctic and back again.

Variety and resurrection
There are some 47 species of terns distributed worldwide, most of them seabirds, although a few breed inland. Typically they have gray and white plumage with a black cap when breeding, although noddy terns reverse this color scheme, and the marsh terns (genus *Chlidonias*) are very dark in breeding plumage.

The Chinese crested tern (*Thalasseus bernsteini*), nicknamed "bird of legend," was presumed extinct until four pairs were discovered in a greater crested tern breeding colony in 2000. The islet hosting this colony is now protected, but the global population still only numbers around 100 birds.

Birdlike figure holds an egg in its hand

Easter Island challenge
Islanders traditionally honored the first person to collect a sooty tern egg from the islet of Motu Nui each year, as commemorated in this stone carving, from around 1900.

SOOTY TERN
Onychoprion fuscatus
This tern is one of the most abundant seabirds; Bird Island in the Seychelles is named because of its presence. It is nicknamed "wide-awake," due to its constant calling.

WHITE TERN
Gygis alba
This beautiful bird of tropical islands is known for laying its egg on a bare branch. Beloved by seafarers, its presence reliably indicates land nearby.

CHAPTER 4

Icy Oceans

The cold waters near the Earth's poles have their own unique set of conditions, and they perform a vital function in regulating the planet's temperature. Life flourishes even in this harsh environment, adapting to freezing temperatures and variable light conditions.

Search party
Ships search for Sir John Franklin's lost 1845 Arctic expedition among the icebergs of Hudson Bay, Canada, in this 1849 illustration.

Ice Shelves and Icebergs

Where the giant ice sheets of the Arctic and Antarctica meet the ocean they continue out across the water, sometimes for hundreds of miles, ending in towering cliffs and breaking off into icebergs.

The Antarctic continent is covered by an ice cap that averages over 6,600 ft (2,000 m) thick and reaches to 15,669 ft (4,776 m) at its thickest point. This vast area locks up around 90 percent of the world's ice and 60 percent of its fresh water. Greenland and the Arctic islands hold most of the rest. Ice extends from these giant ice sheets and spreads out over the sea as ice shelves. These can be huge—Antarctica's Ross Ice Shelf, the world's largest, is about the size of France—and are composed of dense ice that has accumulated from repeated snowfall and the freezing of marine ice over thousands of years. Fresh snow cover is a dozen shades of white, whereas ice cliffs—sheer ice faces at the edge of ice shelves—often appear a vivid blue color, because the exposed deeper layers are more compact and all their tiny white air bubbles have been compressed. Other parts of ice shelves and cliffs are a shimmering green as a result of marine algae frozen into the ice latticework. Floating tidewater glaciers result from mountain glaciers traveling along valleys and reaching the sea. Having ground down rock from the underlying landmass, they can present a color-banded pattern of red, brown, yellow, black, and white.

Ice calving

Ice shelves and glaciers are attached to the ground near the coast, while the outer parts float on the sea. As the floating section moves up and down with the tides and is battered by strong storms, the

Antarctica is the world's **largest desert**

flexure causes great strain, especially where glacial crevices already exist. Huge cracks open up, and chunks of ice break away, crashing into the sea as icebergs—a process known as calving. More than 50,000 large icebergs are calved from Greenland alone each year, and a similar number detach from the Arctic shelves and glaciers. The largest icebergs calve from the Antarctic ice shelves. Warmer air and sea temperatures caused by climate change are accelerating the rate of calving, thinning ice shelves, and raising global sea levels.

Deadly obstacles

Icebergs have been a serious hazard to shipping in the past, and nearly 1,000 collisions have been recorded. The most infamous and tragic of these was the sinking of the *Titanic* on her maiden voyage in 1912 near the Grand Banks off Newfoundland. Ninety percent of an iceberg is below water, so that by the time the crow's nest lookout had spied its tip, the ship was already too near to avoid collision—1,517 people died as a result. More recently, in 2007, the scientific vessel *Explorer* was sunk by an iceberg while sailing near the South Shetland Islands in the Antarctic Ocean. All passengers and crew were saved. Sophisticated iceberg warning systems—using data collected by networks of satellites, aircraft, and ships—now allow safe passage for vessels navigating the far northern and southern polar waters.

Iceberg arch
The action of the wind and waves creates tunnels and holes in icebergs that can connect to form striking, if often short-lived arches, such as this one in the Arctic.

Sea Ice

Every year the seas around the polar regions freeze, creating a 13-million-sq-mile (34-million-sq-km-) blanket of sea ice—the equivalent area of Russia, Australia, and the US combined.

Abandoning ship
The Arctic exploration ship the *Jeannette* trapped in the sea ice north of Siberia in 1881. Having first become ice bound off Wrangel Island in 1879, she drifted 600 miles (1,000 km) in the ice before finally being crushed.

SEA ICE

During the long months of 24-hour darkness in polar regions, sea temperatures plummet to −40°F (−40°C) and sea ice expands across the ocean—when water molecules freeze they occupy more space than they do as liquid. Arctic sea ice extends as far south as Newfoundland and the northern tip of Iceland, while Antarctica's sea ice covers an area twice the size of its continent. Each summer 60–80 percent of this sea ice melts back into the ocean, with no net effect on sealevel. The ice that never melts grows thicker year after year and drifts in the ocean currents, but the area of this permanent sea ice is diminishing quickly as the global climate warms. Arctic summer sea ice coverage is now shrinking by 13 percent each decade. This reduction has a vast

Life on the ice
A polar bear, the largest land carnivore, prowls the frozen sea. It lives on the ice in winter, preying on seals, but will come on to land when the ice melts in summer.

> " The sea, the ice, the snow, it's all changing. You can't travel safely any more. Up and down the coast, it's the same thing. "
>
> DERRICK POTTLE, Inuit hunter, 2018

impact on the lives of Inuit peoples, both practically and culturally, because it forms such an important part of their history.

Formation and destruction

Freezing winds blowing across the ocean surface cause feather-like ice crystals to form. These lock together to create a slushy surface layer known as grease ice that hardens and thickens into a continuous sheet. Wind and waves soon break this into pieces that jostle and bump so that the edges turn up—this is pancake ice. This slowly fuses and thickens into a solid ice cover up to 6½ ft (2 m) thick, which is constantly separating into ice floes and freezing back together again.

As winter sea ice breaks up across the broad Arctic shelves, almost a million marine mammals swim north through the Bering Strait to feed and summer. These summer migrants include narwhals (*Monodon monoceros*), beluga whales (*Delphinapterus leucas*), bowhead whales (*Balaena mysticetus*), walruses (*Odobenus rosmarus*), and polar bears (*Ursus maritimus*). At least 50 million birds migrate to their favored nesting grounds on deltas that fringe the Arctic.

Around Antarctica in spring, the retreating sea ice sparks the largest phytoplankton bloom on the planet. This provides food for small shrimplike krill, which live in large swarms that color the ocean surface a reddish-pink. There are an estimated 700 trillion krill in the waters around Antarctica—the main diet of crabeater seals (*Lobodon carcinophaga*) and a range of penguin and whale species.

Traditional protection
These snow goggles, made from caribou or reindeer antlers, were used by the Thule people, ancestors of the modern Inuit.

Thin slits prevent snow blindness

Marine predators
These aquatic bears are excellent swimmers. They can swim with their eyes open and hold their breath for around two minutes.

Arctic exploration
In this chromolithograph by British artist John Trivett Nettleship, a polar bear rummages through the supplies of an Arctic expedition.

Polar Bear
Ursus maritimus

Polar bears are the world's largest land carnivores, but spend most of their time at sea. These huge predators are perfectly adapted for their freezing environment, but are struggling to survive in a changing climate.

> A newborn **polar bear** only weighs just over **1 lb (0.5 kg)**

Polar bears are the largest species of bear. Males can grow to be twice the size of females, reaching 10 ft (3 m) in length and weighing up to 1,700 lb (800 kg). The Latin name, *Ursus maritimus*, translates to "sea bear." These enormous bears live throughout the Arctic and are well adapted to cope with the bitter cold of their environment. Their thick fur is made up of transparent, hollow hairs that trap heat close to the body. The skin is black to absorb the warmth of the Sun, and the ears are small to prevent heat loss. Clean fur is an excellent insulator, so the bears take baths by rolling in the snow.

Huge paws the size of dinner plates spread the bears' weight over a large area so they do not crash through thin ice. Polar bears are also superb swimmers, using their gigantic paws like paddles to reach swimming speeds of 6 mph (10 kph). They also have outstanding stamina. In 2011, scientists in Alaska tracked a female polar bear swimming continuously for more than nine days, perhaps in a quest to find sea ice.

Polar bears have a powerful sense of smell and they can smell prey—mainly seals—from several miles away. However, although they spend half their time hunting, less than two percent of hunts are successful. Since they have little chance of catching a seal in the water, polar bears have developed hunting strategies to catch a meal. One technique involves waiting patiently above holes in the ice until a seal comes up to breathe,

Koryak pipe
This pipe carved from a walrus tusk was made by the Koryak people of the far east of Russia who live along the coast of the Bering Sea.

Three polar bears walk along the stem of the pipe

HOLLOW FUR

Temperatures in the Arctic can drop lower than –40°F (–40°C) so polar bears need to stay warm. As well as thick blubber, they have two layers of fur for insulation. Dense underfur traps body heat, while longer guard hair is hollow and transparent to capture warmth from the Sun, scatter light, and retain heat. The fur is oily to stop water reaching the skin.

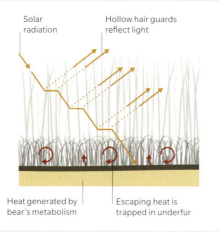

POLAR BEAR FUR AND SKIN

Scientists learn about **polar bears** by taking **DNA** from their **footprints** in the **snow**

> " So powerful an animal as the Polar Bear must necessarily be very dangerous ... "
>
> JOHN GEORGE WOOD, *The Illustrated Natural History Volume 2,* 1865

and then pouncing. Bears also ambush seals resting on the sea ice or sniff out dens where mothers are resting with their pups, and break through the den roof to catch the seals below.

When a female polar bear becomes pregnant, she makes a snow den in which to shelter. Here she fasts for up to eight months, living off her body's fat reserves until the cubs are born and are strong enough to face the outside world. Females typically give birth to two cubs, which stay with their mother for the first 2–3 years of their lives.

Polar bears and people

Polar bears are facing many threats. As the warming climate melts their icy habitat, it is becoming harder for them to hunt their natural seal prey, and they are in danger of starving. Climate change is also increasing wind speeds, which makes it harder for bears to sniff out prey. Drilling for oil and gas in the Arctic can cause harmful oil spills that can damage polar bear habitats.

Inuit cultures have great respect for these huge bears, and some believe they are shape shifters with magical powers. In Inuit mythology, Nanook, the god of polar bears, rules the Arctic and keeps hunters safe on the ice. A Norwegian tale tells the story of a prince cursed to be a white bear by day. If he lives as a bear for a year, the curse will break. But his wife sees his sleeping human form and accidentally drips candle wax on him, waking him, so the curse cannot be broken.

Poster bear
This 1950s poster advertising Anvers (Antwerp) zoo in Belgium used the popular Arctic mammal to draw in visitors.

Mother and cubs
Female polar bears are fiercely protective of their cubs. By the time a male cub is two years old, it may weigh as much as its mother.

POLAR BEAR 177

Mother polar bears stay close to their cubs, hide them from danger, and fight attackers if necessary

Cubs learn to copy their mother's survival skills

Narwhal

Monodon monoceros

In medieval times, seafarers told tales of a sea unicorn, a monstrous fish or sea dragon with a single horn that could impale ships. A real creature was behind these myths—the narwhal.

Mythical beast
This woodcut from the 1500s depicts the fearsome "sea unicorn" with a horn on its head and dragon wings.

As with other "sea monsters," the narwhal's size was often exaggerated in old illustrations. Like its relative the beluga (*Delphinapterus leucas*), it is a small whale, with females reaching 13 ft (4 m) in length, and males up to 16–17 ft (5 m). However, males have an impressive, spiral-twisted horn, which can add another 10 ft (3 m) to their length. The horn is actually a tusk—the left canine tooth, which grows longer as a male calf matures. Although they are classed as toothed whales, narwhals have barely any teeth in their jaws and have to swallow prey whole. Their Latin name, *Monodon monoceros*, means "one tooth, one horn." Narwhals live around the edges of the Arctic Ocean, off the northern coasts of Canada, Greenland, Russia, and Europe. In the dark Arctic winter, they swim and dive deep beneath the sea ice, hunting for

Grouping together
Narwhals often swim in groups or pods of up to 20. This pod is surfacing in a crack in the ice off Baffin Island, Nunavut, Canada.

flatfish, cod, and squid to eat. Like all sea mammals, they need to breathe air, so they rely on cracks or "leads" in the ice where they can surface. In the summer months, as the sea ice melts, narwhals move south into shallower waters, where they give birth to their young, called calves.

Mysterious tusk

For years, scientists were unsure what purpose the tusk served, but they now think it has several uses. Males do not impale prey, but they have been seen side-swiping fish with their tusks to stun them before moving in for the catch. The tusk is packed with millions of sensitive nerve endings that may be able detect water salinity, temperature, and ripples from moving prey. However, since females have no problem finding food, these functions are not essential. In fact, the main purpose of the tusk may simply be to impress females—they seem to prefer males with the biggest tusks as mates.

" The Unicorn is a Sea-beast, having in his Fore-head a very great Horn, wherewith he can penetrate and destroy the ships in his way, and drown multitudes of men. "

OLAUS MAGNUS, *Carta Marina*, 1539

INSIDE THE TUSK

The narwhal's tusk is made up of layers of bone-like material, with softer tissue, blood vessels, and nerves inside. Water washes into the tusk through tiny pores on its surface, allowing the nerves inside to sense it.

CROSS SECTION OF NARWHAL TUSK

- Spiral surface
- Hard dentine
- Nerves running along middle of tusk send signals to the brain
- Soft tissue inside tusk
- Central artery

Braving the ice
A dramatic view of the glacial Arctic Ocean, *Walrus Fishing in Greenland* by François-Auguste Biard (1841) evokes the struggle of Arctic hunters pursuing walruses among the icebergs, floes, and rough waves. Biard painted it after visiting the Arctic in 1839.

Arctic Ocean

The Arctic Ocean basin extends over the North Pole and south to 60° latitude. Unlike the Antarctic, where the South Pole is surrounded by a continent, at the North Pole there is only the icy ocean itself.

The smallest of the world's ocean basins, the Arctic covers an area of about 5.4 million sq miles (14.1 million sq km). It is encircled by the coasts of northern Siberia, northern Scandinavia, Iceland, Greenland, northern Canada, and Alaska.

The Arctic is permanently covered with ice, but this polar ice cap fluctuates during the year. Every winter, the ice spreads to the surrounding shores. In summer, the sea ice melts and retreats

ARCTIC OCEAN

to a much smaller area. Today, global warming has greatly reduced the ice cover during summer. The Arctic is warming faster than anywhere else on Earth. As the white ice (which reflects energy from Earth into space) retreats and the darker water (which absorbs solar energy) increases, this exacerbates the warming effect.

This loss of sea ice may affect all the life forms in this habitat. Phytoplankton may increase due to the extra sunlight, but other organisms at the base of the food chain, such as zooplankton, may be adversely affected, which will then impact the fish, birds, and other species that feed on them. The loss of sea ice will reduce the habitat for seals and walruses, which live and breed on the ice, and for polar bears. This is highly stressful for the animals, and can be dangerous for people if polar bears enter human settlements in search of food.

Peoples of the North

Humans have lived around the Arctic Ocean for thousands of years. They include the Inuit peoples of North America, the Sámi of northern Europe, and the Nenets in Russia. These cultures developed skills for sea ice fishing, kayaking, and dogsledding that date back at least 9,000 years. Some of these traditions still continue today, albeit with the help of modern technology.

Western explorers tried for centuries to find sea routes through the Arctic between the north Atlantic and the Pacific, only succeeding in the 1800s. Today, there are concerns that the warming Arctic will see increased marine traffic, fishing, and mining. To protect it, in 2015 the US, Canada, Russia, Greenland (Denmark), and Norway signed an agreement to avoid fishing around the North Pole.

Ice bear
The polar bear (*Ursus maritimus*) relies on the floating sea ice to rest, travel, and give birth on, as well as swimming long distances. Where ice and water meet, polar bears hunt their favorite prey—the ringed seal (*Pusa hispida*).

> "**Arctic**" means "**of the bear**," and relates to the **Great bear constellation** visible in northern skies

Living in cold water
Most sharks prefer to live in warmer waters, but the Greenland shark (*Somniosus microcephalus*) is found in the Arctic and the far north of the Atlantic. This slow-swimming, slow-growing species is thought to live for up to 400 years.

Shrimplike parasites attach to the eyeballs of Greenland sharks

The frequency of tail beats and the swimming speed are very low for a fish of this size: just under 1.9 mph (3 kph)

Walrus

Odobenus rosmarus

With a scientific name that means "tooth-walking sea horse," these large Arctic mammals are known for their huge tusks, which they use to move around, find food, and create holes in the ice.

Humanoid figure has exaggerated features

Indigenous carving
This carved walrus ivory figure from the Thule people of northwest Greenland was said to protect its owner against enemies.

Walruses live in Arctic and sub-Arctic conditions where they forage for food in shallow water. Their diet consists mainly of clams, shellfish, and snails, and they remove the flesh from their prey by creating a vacuum with their tongue. Walruses spend two-thirds of their life in the water, including giving birth and sleeping. They migrate to stay close to the sea ice, climbing out in groups known as "haul-outs" to rest. As the warming climate causes sea ice to melt, walruses are losing their habitat and are forced to haul out on land or migrate to unfamiliar places. Male walruses are around one third larger than females, and can grow up to 12 ft (3.7 m) in length. Both male and female walruses have tusks, which are long canine teeth. They dig them into the ice to haul their huge bodies out of the water. During the mating season, tusks act as weapons for fighting males, and can also be used to also protect against predators—polar bears and orcas. In the 16th century, people believed that walruses

would hang off the cliffside by their tusks while sleeping—making them easy for hunters to catch.

Walrus hunting
Indigenous peoples have hunted walruses for thousands of years. They eat the meat, carve tusks into tools, use the blubber for fuel, and use the hides to cover boats. Commercial hunting during 17–19th centuries caused populations to dwindle so much that the walrus became a protected species in 1952.

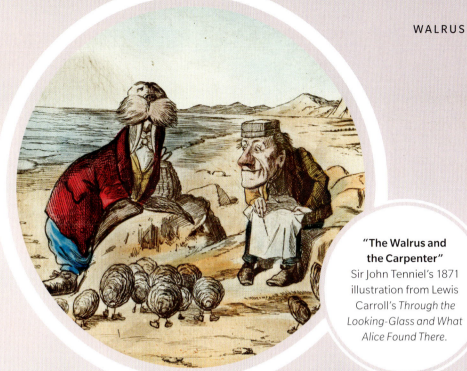

"The Walrus and the Carpenter"
Sir John Tenniel's 1871 illustration from Lewis Carroll's *Through the Looking-Glass and What Alice Found There.*

> "'The time has come,' the Walrus said, 'To talk of many things: Of shoes—and ships—and sealing-wax—Of cabbages—and kings—'"
>
> LEWIS CARROLL, "The Walrus and the Carpenter," 1871

Living in groups
Walruses are social animals that huddle in large groups, but fights and stampedes often break out.

Hauling ashore
In 1916, British explorer Ernest Shackleton had to escape when sea ice crushed his ship; this image shows him and his men landing at South Georgia after sailing their lifeboat 800 miles (1,300 km) across the ocean.

Southern Ocean

Oceanic waters completely surround the continent of Antarctica. Unlike other ocean regions, it has no defined boundary, but it forms a ring around the southernmost latitudes of Earth. It is known for its wild waves and storms, and for its wildlife.

Wild world
Deep, cold Antarctic waters contain large amounts of nutrients and dissolved oxygen, supporting a diverse ecosystem of sea stars, feather stars, sea slugs, sea cucumbers, sponges, corals, and sea spiders.

At 8,478,800 sq miles (21.96 million sq km), the Southern Ocean is Earth's second-smallest ocean. It is also the world's youngest ocean, having formed when Antarctica separated from South America around 34 million years ago. Its northern boundary is generally considered to be latitude 60° south, and its southern boundary is the continent of Antarctica.

The thick ice cap that builds up on the Antarctic continent gradually flows outward into the sea, forming huge floating ice shelves that tower up to 160 ft (50 m) above the water surface. As sections of the ice crack and break away, new icebergs form, or "calve." Some Antarctic icebergs are enormous; the A-76 iceberg, first observed in 2021, measured 110 miles (105 km) long. The US National Ice Center tracks large icebergs by satellite as they float out into the ocean, so ships can avoid them.

The Southern Ocean plays a pivotal role in the circulation of water through the oceans. It connects with, and shares waters with, the South Atlantic, South Pacific, and Indian Oceans. Its own complex pattern of currents includes the world's longest current: the Antarctic Circumpolar Current, which travels continually eastward around the continent. In addition, seawater chilled by winds off Antarctica sinks to the bottom of the ocean and flows north. At a point called the Antarctic Convergence, this cold water meets warmer water from the three connecting oceans. The mixing and up-welling of currents in this zone stirs up high levels of oxygen and nutrients that nourish abundant phytoplankton and krill. These organisms, in

Trunklike, inflatable snout amplifies the male's roars when it challenges other males

Southern elephant seal
The world's largest seal is the southern elephant seal (*Mirounga leonina*) that inhabits the islands off the coast of Antarctica.

turn, form the basis of food chains that include animal species from worms, mollusks, crustaceans, and fish to birds, seals, and whales.

Antarctic wildlife

Despite the polar conditions, Antarctica supports a wide diversity of animals on the land, in the ocean, and on the ice. Some species are found nowhere else on Earth, such as the notothenioids, a group of fish that have antifreeze glycoproteins in their blood. Birds such as albatrosses nest on the rocky ground of Antarctica. Seals and whales feed in the icy water. Perhaps the animals most closely associated with Antarctica, however, are penguins (see pp.188–191). They include the emperor penguin (*Aptenodytes forsteri*), the largest penguin species; it breeds during the Antarctic winter, with parent birds walking up to 75 miles (120 km) between their nest sites and the ocean.

" We were a tiny speck in the vast vista of the sea—the ocean that is open to all and merciful to none. "

ERNEST SHACKLETON,
South: The Story of Shackleton's Last Expedition 1914–1917, 1919

At home in the ice
Four crabeater seals (*Lobodon carcinophaga*) circle an iceberg beneath the waves of the Southern Ocean. Crabeaters are the world's most common seal species with millions of individuals living on the pack ice that floats in the southern seas and feeding on plentiful Antarctic krill (*Euphausia superba*).

Penguins

Spheniscidae

For the penguins, with the evolutionary sacrifice of flight came an unparalleled ability to swim. Few other birds have adapted so completely to a life on and in the water.

Phillip Island's coastline is home to conservation areas for seabirds including penguins and gulls

A little adventure
Phillip Island offers tourists the chance to watch Australian little penguins (*Eudyptula novaehollandiae*) as they come ashore at dusk.

There are 18 species of penguin. Their closest living cousins are albatrosses and their relatives (see pp.144–145), but the two lineages parted ways about 70 million years ago and have since evolved in very different directions. Penguins' wings are now sturdy flippers that do not allow flight, but enable powerful, rapid, and agile swimming underwater. With short legs positioned so far back on their bodies that they can only stand upright or lie flat on their bellies, penguins can appear awkward on land, but they paddle strongly with their feet on the water's surface, switching to wing-propelled swimming once underwater.

Penguins come to land to nest. They need safe places to leave the water, and most species can manage long treks, athletic leaps, or both. Their counterparts in the northern hemisphere, the auks (Alcidae), have retained the power of flight, and nest on cliffs out of reach of predators. The extinct great auk (*Pinguinus impennis*), however, was flightless, and it is believed that when Western

Dark feathers help hide the penguin from predators when viewed from above in the water

explorers saw what are now called penguins for the first time, they named them "penguin" based on the great auk's species name.

Location significance
Being flightless limits where penguins can breed, but they are widespread across the southern hemisphere. The Galápagos penguin (*Spheniscus mendiculus*) inhabits several islands of the archipelago in the northern hemisphere. However, penguins are best known as inhabitants of Antarctica, with the emperor penguin (*Aptenodytes forsteri*) being the only bird to breed during winter in Antarctica. Adélie penguins (*Pygoscelis adeliae*) also breed in Antarctica, with the chinstrap (*Pygoscelis antarcticus*), gentoo (*Pygoscelis papua*), and macaroni penguins (*Eudyptes chrysolophus*) having small breeding colonies on the continent. Between them, these penguins form a total breeding population of about 20 million pairs. Others breed farther north on subantarctic islands and the coastlines of South America, Africa, and parts of Australasia.

Penguins are colony-nesters, like most seabirds. Because they find their food in the open sea, they have no need to defend a feeding territory on land, and there are benefits to nesting in a group, such as shared vigilance against predators and, in some cases, warmth. Adult penguins have few enemies on land, but

The pigment spheniscin is what gives many penguins their yellow color

" I find penguins at present the only comfort in life ... one can't be angry when one looks at a penguin. "

JOHN RUSKIN, Letter to Charles Eliot Norton, 1860

Out of the darkness
As the permanent night of Antarctic winter ends and the days begin to grow brighter, young emperor penguins grow up fast under the care of their parents.

Dense, fur-like plumage and a thick layer of body fat keeps emperor penguins warm in their icy and windy habitat

AQUATIC ADAPTATIONS

The penguin skeleton has several adaptations for swimming and diving. Penguin bones lack the hollow spaces found in flying birds' bones. This increases the body's density, resulting in faster and deeper dives. The leg position gives a streamlined shape while swimming underwater and provides power for surface swimming.

PENGUIN SKELETON

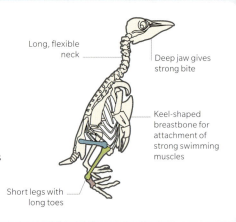

Long, flexible neck
Deep jaw gives strong bite
Keel-shaped breastbone for attachment of strong swimming muscles
Short legs with long toes

Penguins can stay underwater longer than any other bird. This, coupled with their superb swimming skills and dense, heavy bodies, enables them to make deep dives. One emperor penguin was recorded diving to 1,850 ft (564 m), and this species routinely dives to 650 ft (200 m). The blood of an emperor penguin has a high hemoglobin content to hold extra oxygen. It uses oxygen efficiently while diving and slows its metabolism to improve energy efficiency. During a dive, an emperor penguin can swim at 7 mph (11 kph), but the smaller gentoo reaches up to 22 mph (36 kph)—faster than most seals and sea lions.

Penguins spend the nonbreeding season at sea. When rearing chicks, they swallow fish then regurgitate it to their young. As the chicks grow, they often form large crèches to stay safe from predators while the parents are away feeding. A returning parent is harassed by every chick it walks past, but it homes in on the call of its own

various birds and mammals take their eggs and chicks, so adults birds are vigorous in nest defense. Introduced predators such as stoats and ferrets can also cause havoc. Yellow-eyed penguins (*Megadyptes antipodes*) in New Zealand have suffered serious decline due to this and other causes.

Emperor penguins can dive under the surface for up to 20 minutes

Tropical outlier
The Galápagos penguin has the most northerly distribution of any species, and when breeding it swims in equatorial waters.

offspring. If a chick dies, its parents may adopt (or attempt to adopt) another. Strong parenting instinct is matched by the penguin pair bond, with pairs usually reforming each year.

In myth and culture

Maori mythology talks about the yellow-eyed penguin, and the Fiordland penguin (*Eudyptes pachyrhynchus*) or "tawaki" of New Zealand's South Island. The latter is said to be the incarnation of a storm god, wearing bolts of lightning in the form of a vivid yellow crest above each eye.

The crested macaroni penguin of South America takes its name from an 18th-century nickname for a well-groomed man, and penguins are celebrated for their black-and-white coloring, which resembles formal wear. A black upper side with a white belly is common across many seabirds, and is a form of camouflage—they are less visible in water when viewed both from above and from below.

> **Same-sex pairing** has been noted in some penguin **species**

New Zealand giant penguins
Fossils show that some ancient penguins were much larger than modern species. *Pachydyptes ponderosus* of New Zealand (right) measured up to 4⅓ ft (1.3 m) tall, while the Australian *Palaeeudyptes klekowskii* stood more than 6½ ft (2 m) tall.

GENTOO PENGUIN
Pygoscelis papua
The fastest swimming bird in the world, gentoos accelerate through the water angling their wings to cut water resistance. They use their speed to avoid orcas and leopard seals.

CHINSTRAP PENGUIN
Pygoscelis antarctica
With a population of around 8 million, the chinstrap is one of the most abundant species. It breeds on subantarctic islands and feeds on fish, krill, and shrimp.

SOUTHERN ROCKHOPPER PENGUIN
Eudyptes chrysocome
The smallest of the crested penguins, this species is a skilled leaper, using its wings for balance as it bounds along.

Leopard Seal
Hydrurga leptonyx

Also known as the "sea leopard," the leopard seal gets its name from its sleek, spotted coat. This Antarctic predator is leopard-like in other ways too, with huge jaws and teeth, and a fast, ferocious hunting style.

Teeth are used for both hunting and filter-feeding

Skull profile
Leopard seals have large skulls, which can be up to 16 in (42 cm) long.

Found mostly in the Antarctic, the leopard seal is the second-largest of all seals, after the enormous Southern elephant seal (*Mirounga leonina*), and has a long, streamlined body. Females are bigger than males, reaching 12 ft (3.7 m) in length.

Underwater, leopard seals' muscular bodies and front flippers give them both impressive speed and steering power. They can swim through the water at up to 25 mph (40 kph), flipping and turning in a split second. These skills are essential for chasing and catching their main prey—fast-moving penguins. They also hunt other seabirds, squid, fish, and even smaller species of seal.

On ice or on land, however, they are slower, because their relatively small forelimbs are not powerful enough to help them move. Instead, they wriggle and lunge forward. Although this can still enable relatively speedy movements, most potential prey, such as penguins, can usually escape to safety.

Powerful jaws and teeth
Inside the leopard seal's large skull are huge jaws with a massive gape—bigger than a human's head. The mouth is upturned at the sides, seeming to give the seal a deceptively friendly smile. At the front are long, sharp canine teeth, similar to a bear's or a big cat's, adapted to gripping and tearing fast, slippery prey. Farther back are rows of interlocking teeth that sift smaller species from the water. Leopard seals, especially when younger, use this method to feed on tiny, shrimplike krill that swarm the Antarctic Ocean in their trillions.

On the ice
Leopard seals prefer to hunt among the floating Antarctic ice, using the ice floes to rest on, and to avoid their only natural predator—the orca.

Curious creatures
A large female leopard seal, drifting on Antarctic sea ice rears up to inspect the photographer. Females especially have been observed showing curiosity and even playfulness toward humans—but they can also be dangerous.

They sometimes hunt by patrolling the edges of the ice just below the surface, waiting for penguins to jump in. Like other polar seals, females give birth on the ice, where they feed and care for their single pup. The mother scrapes a hollow as a nest for her baby.

Leopard seals are solitary creatures, and only meet up to mate. To attract a female, the male hangs upside down in the water, and sings his own unique song.

Hunter and prey
This illustration taken from a zoology book from 1839 depicts a leopard seal and several species of the seabirds it feeds on.

Eyes are mounted on the side of the head, unlike those of most seals

A **leopard seal's** head is similar in size to a **grizzly bear's**

CHAPTER 5

Beneath the Waves

The ocean depths are one of Earth's last true frontiers. Conditions beneath the surface range from the warm waters and bustling marine life of the upper layers to the crushing pressure and pitch darkness of the deepest ocean.

Sunlit Zone

Life flourishes in this, the uppermost ocean zone, with a mixture of apex predators, roaming schools of fish, and trillions of plankton, as well as undersea forests of plants and algae.

> The sunlit zone comprises about 3 percent of Earth's ocean

The sunlit zone, or euphotic zone, comprises the layer of the ocean that lies between the surface and depths of up to 650 ft (200 m). It is both relatively the warmest and the lightest of the ocean zones (see pp.28–29) because of the presence of sunlight. Aristotle (384–322 BCE) described many marine species of the sunlit zone in one of the world's first marine biology texts, *History of Animals*. However, scientists now know that he had barely scratched the surface of the life forms that can be found in this zone. The whole spectrum of ocean life throughout the globe depends on the sunlit zone. As it is bathed in sunlight, it

Sunlit living
Stingrays, reef sharks, and butterfly fish swim in shallow waters over coral reefs in Bora Bora, French Polynesia.

Slugs of the sunlit zone
Nudibranchs are shell-less marine gastropods. These *Godiva quadricolor* in the Mediterranean Sea are known for their bright colors.

> " The sea, once it casts its spell, holds one in its net of wonder forever "
>
> JACQUES COUSTEAU, *Life and Death in a Coral Sea*, 1971

is home to microscopic organisms called phytoplankton that generate their energy through photosynthesis and which form the basis of the ocean's ecosystem.

The ocean food chain

There are two main types of phytoplankton: microscopic algae and cyanobacteria. The algae bloom in huge numbers: a bucketful of water from the sea surface can contain up to 15 million of them. An estimated 100 trillion trillion cyanobacteria live across the ocean; in total, some 6,600 billion tons (6,000 billion metric tons) of photosynthetic plankton grow each year (see pp.198–199). This phytoplankton blooms in the spring, and microscopic zooplankton feed on it. These are preyed on by copepods—tiny elongate crustaceans about the size of a sand grain with waving antennae. There can be as many as 2,830 of these in one cubic foot of ocean water (100,000 per cubic meter). A multitude of species of fish, seabirds, whales, and dolphins in turn feed on the copepods.

Sunlit habitats

Coral reefs (see pp.204–207), kelp forests (see pp.118–119), coastal wetlands, and open-ocean islands of *Sargassum* seaweed (see pp.158–159) are all important habitats found within the sunlit zone. Kelp forests in particular, which grow in cool temperate seas, are among the most productive marine ecosystems on the planet. Giant kelp (*Macrocystis pyrifera*) can grow some 20 in (50 cm) in a single day and can grow to more than 100 ft (30 m). This nutrient-rich environment not only provides an abundant and stable supply of food, but also shelter and protection for a wide range of animals, from invertebrates to marine mammals.

Sea creatures
The 15th-century painting shows the fictional story of how Alexander the Great tried to seize the ocean. He is lowered into the water, which is teeming with life.

PLANKTON

Depicted in detail
In the 19th century, German biologist Ernst Haeckel created intricate illustrations of plankton, such as this.

The word plankton comes from the Greek planktos, which means to drift or wander

Plankton

Plankton are a variety of marine organisms that are carried along by the current—although some can swim. There are more plankton in the ocean than stars in the sky, and plankton blooms can be seen from space.

The plankton contains many different types of lifeforms. It includes microscopic organisms such as diatoms (Bacillariophyceae), a form of single-celled algae, as well as larger animals such as the enormous lion's mane jellyfish (*Cyanea capillata*). Although the latter can reach lengths of 6½ ft (2 m), it is still unable to swim against the movement of the ocean.

There are two main types of plankton. Phytoplankton are bacterial and plant matter, including single-celled organisms such as diatoms and dinoflagellates (Dinoflagellata). Zooplankton are animals, including krill (Euphausiacea), shrimplike crustaceans, and salp (Salpidae), small invertebrates known as "sea grapes." These organisms feed many other animals, so without them, the entire ocean food web would collapse.

Humans have known about plankton for centuries: in 1673, Dutch microbiologist Antonie van Leeuwenhoek first saw plankton through a basic microscope. However, it took more than 200 years for the word "plankton" to be coined by German biologist Victor Hensen in 1887.

As the ocean warms, changes are occurring in plankton abundance and location—different species prefer different temperatures. Because plankton is such an important food source, it is believed this could have wide-ranging effects on the global marine food chain.

Satellite swirls
This enhanced NASA image captures huge swirls of phytoplankton in a nutrient-rich area of the Atlantic Ocean near South America.

COMB JELLY
Ctenophora

Comb jellies' gelatinous bodies have tiny hairs called cilia, or combs, which flicker in rainbow colors to help them move through the water.

COPEPOD
Copepoda

These minuscule plankton are only 1–2 mm long, but some species can swim distances of 300 ft (90 m) in just one hour.

DINOFLAGELLATE
Dinoflagellata

Dinoflagellates are a type of phytoplankton. These single-celled algae underpin many marine food webs.

Undersea study

Types of fish, anemones, seahorses, lobsters, starfish, barnacles, and seaweed are among the 75 species on display in artist Christian Schussele's scene of undersea life. Dating to 1859, the watercolor was designed to accompany the pamphlet *Ocean Life*, published by Philadelphia physician and amateur naturalist James M. Sommerville. This was one of several works that reflected a growing interest in marine life in the mid-19th century, prompted by the laying of the first transatlantic telegraph cables.

Seaweeds

Seaweed is the common name for around 12,000 species of plantlike organisms that live in the oceans and photosynthesize. They have no roots, instead transporting nutrients through their bodies by diffusion.

Seaweeds are large, multicellular oceanic algae that are divided into three groups based loosely on their photosynthetic pigment color: green, red, and brown. The majority of seaweeds have a much simpler structure than land plants. They lack roots, stems, and leaves, and instead have a holdfast that latches them on to their substratum, a stemlike section called the stipe, and flat structures known as blades. Many species also have air-filled bladders to keep them reaching up toward the light.

There is a vast diversity of seaweeds—some are barely visible to the naked eye, while others grow into enormous underwater forests that are some of the most productive habitats on Earth.

Food and medicine

Humans have been foraging seaweed for millennia. Many species of seaweed are edible and they remain an important part of the diet in east Asia and other parts of the world.

Many cultures have also used seaweeds for medicinal purposes. The earliest written reference to this comes from the *Epic of Gilgamesh* (c. 2100 BCE), which tells how the hero Gilgamesh dived to the bottom of the sea in search of a youth-giving seaweed. Later, the Romans used seaweeds to treat wounds, burns, and rashes. There is even some evidence that ancient Egyptians may have used them to treat tumors. Today, seaweeds are used in a vast range of products, from cosmetics to animal feed and fertilizer.

However, the importance of seaweeds extends much further. They provide beautiful, highly biodiverse habitats and nursery areas for fish and other marine animals; protect our coastlines from wave energy; filter harmful pollutants; and improve water quality. They also provide habitats, food, and breeding grounds for countless other species, including fish, turtles, birds, and crabs.

RED CORALLINE ALGAE
Corallinales sp.

Most seaweeds have soft tissues, but coralline algae incorporate calcium carbonate into their cell walls, making them hard and unpalatable to most seaweed grazers.

SEA LETTUCE
Ulva lactuca

This common seaweed with bright-green, ruffled fronds resembles lettuce leaves. The fronds are almost translucent, because they are just two cells thick.

PIGMENT POSITIONING

A seaweed's pigments dictate where it can survive: each type of pigment absorbs different wavelengths of light, which are available at varying depths. Red seaweeds live in the deepest waters, green at the shallowest, and brown in between.

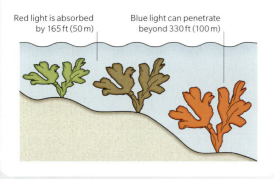

Red light is absorbed by 165 ft (50 m)

Blue light can penetrate beyond 330 ft (100 m)

Giant kelp
Growing to more than 160 ft (50 m) in length, giant kelp (*Macrocystis pyrifera*) is the world's largest seaweed.

Fronds can grow 24 in (60 cm) a day

Dazzling diversity
Seaweeds come in an array of colors and forms, as this lithograph by French artist Adolphe Philippe Millot (1857–1921) illustrates.

Teeming with life
Hard coral forms a basis for soft corals and a place for fish to feed and hide from predators, as in this Pacific reef off the coast of Fiji.

Mountainous star coral spawning
Corals spawn at night, often at full moon, with polyps simultaneously releasing millions of egg and sperm bundles into the water.

Coral Reefs

Sometimes called the rainforests of the ocean, coral reefs are the most diverse of all ecosystems. Although they occupy less than 1 percent of the ocean floor, they support around 25 percent of all marine species.

Reefs can grow into large, towering structures, but they are created by minuscule creatures: coral polyps, which are related to jellyfish and sea anemones. Some species of coral polyp are solitary, but many create vast colonies containing millions of individuals.

The world's largest coral reef system is the Great Barrier Reef off the northeastern coast of Australia (see pp.210–211), which covers 133,000 sq miles (344,400 sq km) of seafloor and includes more than 2,900 individual reefs. Southeast Asia's Coral Triangle covers 2.3 million sq miles (6 million sq km) of coastal seafloor in the Pacific and is recognized as a global center of marine biodiversity, with 76 percent of the world's coral species.

Reef-forming corals
There are more than 800 species of coral, but not all of them are reef-builders. Those that form reefs are called hard corals, because of their ability to extract calcium carbonate from seawater and transform it into a hard external skeleton. These skeletons grow by about ¼ in (0.5 cm) per year.

Sunlit waters
Tropical reef-building corals thrive in shallow, warm waters, like this table coral growing just off Gaafu Alifu Atoll, Maldives, Indian Ocean.

> " Such formations surely rank amongst the
> wonderful objects of this world ... "
>
> CHARLES DARWIN, 1836

BUILDING A REEF
When a coral larva finds a suitable hard structure on which to settle, it transforms into a coral polyp. The polyp extracts calcium carbonate from the water to build itself a cuplike skeleton called a corallite. Slowly, the polyp grows, laying down more layers of skeleton and producing more polyps, connected to each other by a thin layer of tissue. Over decades and even centuries, the coral colony expands upward and outward.

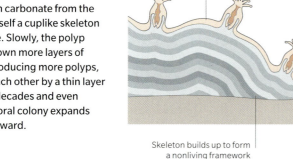

CORAL REEF

Each coral polyp has a ring of stinging tentacles around the mouth, enabling it to catch food such as plankton and tiny fish. However, many coral polyps derive more than 90 percent of their nutrients from their symbiotic relationship with microscopic algae called zooxanthellae that live inside their tissues. These algae photosynthesize, chemically converting sunlight into sugars, and in so doing they provide the coral polyp with food and oxygen. The polyps provide waste carbon dioxide, which the zooxanthellae use for photosynthesis, as well as providing a habitat.

While corals exist throughout the ocean, reef-forming corals can both live in shallow waters in the tropics, where there is plenty of sunlight for symbiotic zooxanthellae, or form vast deep-sea coral reefs in permanent darkness.

Once hard corals have laid the foundations of a tropical reef, it is populated by soft corals, sponges, and algae. As corals and other animals die or are eaten, the debris forms sand, which further stabilizes the reef. Parrotfish (subfamily Scarinae) are noted for eating coral and excreting the hard, calcium-rich waste as sand. As polyps die, new ones grow on top, building the reef up over time.

British naturalist Charles Darwin first identified the three main types of reef in 1842, following his visit to the Indian Ocean during his round-the-world voyage on HMS *Beagle*. These classical forms were fringing reefs, which develop along coastlines; barrier reefs, separated from the coast by a wide lagoon; and coral atolls, which develop on the flanks of volcanic islands. These terms are still in use today, although many scientists now also include other types.

A delicate environment
The first corals evolved about 500 million years ago, but reef-building species similar to modern corals first evolved around 60 million years ago. Most of today's reefs first formed about 10,000 years ago, after the end of the last Ice Age.

Creation chant
This mural by Hawaiian artist Kahi Ching is inspired by the ancient Hawaiian creation chant "Kumulipo," in which the coral polyp (*ko'a*) is portrayed as one of the first life forms to be born during the cosmic night when the world came into being.

However, corals are sensitive to changes in the seawater. If the seabed sinks or sea levels rise, they can be killed when they can no longer photosynthesize. If the water becomes suddenly warmer or colder, or is polluted by excess nutrients, they become stressed and expel their zooxanthellae, which turns them white (see p.211). Changes in ocean chemistry, notably acidification due to rising levels of carbon dioxide, are making the problem worse. Around half of all tropical reefs have been lost in the last century, and scientists believe that by 2050 up to 90 percent may have vanished.

Reefs and humans

For millennia, Indigenous peoples have venerated reefs for spiritual as well as practical reasons. The Yidinji people of the Queensland coast have a legend in which the Great Barrier Reef was formed when two brothers angered Bhiral, the Creator. Bhiral threw lava from the sky into the ocean, causing the sea level to rise; when the lava cooled, it formed the base for the reef. Scientific evidence of volcanic activity about 7,000 years ago supports this account.

Today, many people depend on reefs for fishing, for protection from the open sea, and for the revenue tourism brings. Reef plants and animals are being researched as sources of treatments for conditions including cancer and heart disease. To protect such unique habitats, the United Nations and many countries have designated tropical reefs as protected areas.

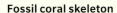

Septa (walls) radiate out from the center of the polyp

Fossil coral skeleton
This fossil coral belongs to the order Scleractinia. These corals evolved in the Triassic period (252–201 million years ago); their descendants are the corals that exist today.

Paternal instincts
A male seahorse has developed a muscle and bone structure that allows him to open the pouch and release his young, as shown in this illustration.

Small pectoral fins on the side of the head help with steering and stability

Toothless, tubelike snout sucks up small crustaceans

Seahorses

Hippocampus

Seahorses have captured the human imagination for thousands of years. They belong to the genus Hippocampus, a name that translates as "horse sea monster." There are more than 45 species of seahorse, inhabiting all but the coldest coasts worldwide.

Seahorses can be found in warm and temperate salt water shallows around the world. In the 1800s, they were mistaken for insects because their entire bodies are covered with a protective, bony armor rather than fish scales. Their prehensile tails wrap tightly around objects such as coral or seaweed to prevent seahorses from being swept away in strong ocean currents. They swim upright and travel at a top speed of just 5 ft (1.5 m) per hour, making them one of the slowest-moving fish in the ocean.

Blending in
What they lack in speed, seahorses make up for with camouflage. They have specialized pigment structures in their skin called chromatophores, which allow them to change color to match their surroundings. By blending into the background, seahorses are able to ambush prey while escaping the sharp eyes of predators. This special adaptation is also used in communication and courtship where it is believed to help reinforce the bond between a male and female.

The most remarkable characteristic of seahorses is that the male carries and gives birth to the young—an adaptation unique to seahorses and their close relatives, the sea dragons. After an elaborate and often lengthy courtship dance, the female seahorse deposits her eggs into a kangaroo-like pouch on the male's abdomen—the brood pouch—where he fertilizes them. Two to four weeks later, the male's body begins to contract violently, forcing out as many as 1,000 miniature, fully formed baby seahorses. The contractions can last up to 12 hours. As

Unusual shape
Seahorses share their scientific name—Hippocampus—with part of vertebrates' brains. It is thought that scholars gave it this name because its shape resembles a seahorse.

SEAHORSES

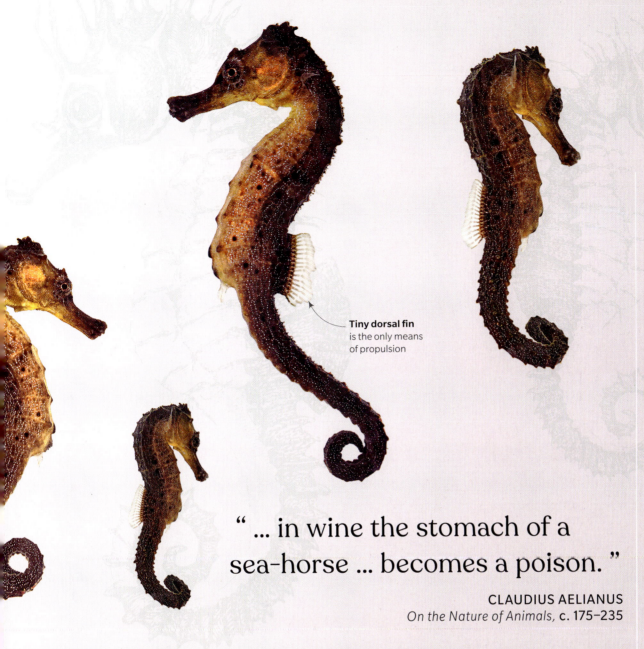

Tiny dorsal fin is the only means of propulsion

" ... in wine the stomach of a sea-horse ... becomes a poison. "

CLAUDIUS AELIANUS
On the Nature of Animals, c. 175–235

PYGMY SEAHORSE
Hippocampus bargibanti

This tiny seahorse is ¾ in (2 cm) long and blends in with the colorful sea fans on which it lives in the Western Pacific.

LONG-SNOUTED SEAHORSE
Hippocampus guttulatus

Also known as the spiny seahorse, this species is recognized by its long snout and the fleshy "mane" along its back, and is one of just two species found in UK waters.

BIG-BELLY SEAHORSE
Hippocampus abdominalis

This is the largest seahorse species in the world at 14 in (35 cm) and is native to Australia and New Zealand.

soon as they are born, the young must fend for themselves, and most of these tiny fry become food for other marine creatures. Fewer than one percent survive into adulthood.

Seahorses and people

The first known seahorse images, found on cave walls in Australia, are 6,000 years old. Sketches of seahorses appear in burial chambers built by the Phoenicians and Etruscans, and the coffin of one Egyptian mummy has a seahorse painted on it. The ancient Greeks and Romans believed seahorses could cure ailments such as leprosy and rabies. Seahorses are still used in some traditional medicine practices. Today, seahorses' numbers are declining due to non-selective fishing, habitat loss, invasive species, and climate change.

Neptune's chariot
In Roman mythology, Neptune's chariot was pulled by a beast that was half-horse, half-fish. The creature was called Hippocampus, the genus name for seahorses today.

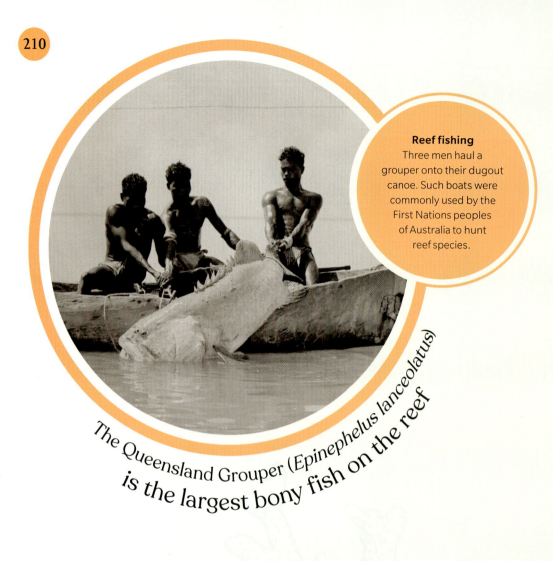

Reef fishing
Three men haul a grouper onto their dugout canoe. Such boats were commonly used by the First Nations peoples of Australia to hunt reef species.

The Queensland Grouper (Epinephelus lanceolatus) is the largest bony fish on the reef

Great Barrier Reef

Lying off the northeastern coast of Australia, the Great Barrier Reef is not only the world's largest coral reef system, but also its largest living structure.

The Great Barrier Reef stretches more than 1,400 miles (2,300 km) along the coast of Australia and covers more than 133,000 sq miles (344,000 sq km). This vast network of interconnected reefs and islands is among the most biodiverse habitats on Earth and has been designated one of the seven wonders of the natural world. It is home to 25 percent of all known marine species, including 1,500 types of fish and more than 600 species of coral, many of which are found nowhere else.

The Great Barrier Reef first started to form about 20 million years ago. When sea levels fell, the developing reef was exposed and growth halted, only to be re-submerged and start growing again when waters rose. After the last ice age, the ocean flooded in over the continental shelf once more and the reef that exists today began taking shape. Every reef is formed from the skeletons of coral polyps, some just a few millimeters across. Over millions of years, the structures formed by these tiny components can grow up to 1,640 ft (500 m) high.

The lives of many First Nations peoples have been deeply entwined with the Great Barrier Reef for around 60,000 years. Dreamtime stories that have been passed down the generations contain knowledge of the reef and its spiritual significance. One story tells of a time before the

Massive mosaic
The Great Barrier Reef Marine Park is made up of at least 3,000 reefs and 980 islands divided by narrow passages. In total, it covers an area about the size of Japan.

land flooded when people lived where the reef now stands. This matches up with recent scientific evidence of sea level changes.

Uncertain future

The reef plays a vital role in coastal protection, dampening the power of waves, storms, and floods. It also contributes billions to the Australian economy each year through tourism and fishing. However, its future hangs in the balance. Climate change is a major threat, and poor water quality from sediment run-off and pollution, storms, and outbreaks of coral-eating starfish also put it under strain.

GHOST REEF

Corals get their color from zooxanthellae, the microscopic algae cells that live within their tissues and provide them with food. When they become stressed, corals expel their zooxanthellae, turning a ghostly white. This is called coral bleaching. Without their algal helpers, the corals begin to starve and, if the stress is not removed, eventually die. A major cause is stress from higher water temperatures caused by climate change.

BLEACHED CORAL

BENEATH THE WAVES

> **Leafy sea dragons** are South Australia's **official state fish**

Cunning disguise
Leafy sea dragons are among the most intricately camouflaged creatures on Earth. They propel themselves through the water using fins that are shaped like seaweed.

Leafy Sea Dragon

Phycodurus eques

Sea dragons are close relatives of seahorses and are found only in Australian waters. Of the three known species, the leafy sea dragon is one of the world's strangest looking fish.

Although they bear some resemblance to the fire-breathing dragons of mythology, sea dragons are in fact tiny, toothless fish that specialize in staying out of sight. Like seahorses, they are weak swimmers, so camouflage is key to their survival. Their bodies are elaborately adorned with delicate, leafy appendages, which make them practically invisible among the kelp and seagrass of the rocky reefs on which they live. This is reflected in their scientific name, which means "seaweed skin." Sea dragons can also change color to match their surroundings and even imitate the swaying motion of seaweed by rocking back and forth.

Leaf-shaped appendages cover the entire body to provide an effective camouflage

Leafy sea dragons lack the grasping tails that are characteristic of seahorses (see pp.208–209). Instead, they drift gently in the current, sucking up thousands of mysid shrimp and other plankton through their tubelike snouts.

Like seahorses, it is the male sea dragon who gives birth. The female deposits hundreds of eggs onto a brood patch on the underside of the male's tail, where he fertilizes them. Each bright pink egg is enclosed in a blood-rich cup within the tissues of the brood pouch. The blood vessels deliver oxygen to sustain the developing embryos. A few weeks later, the male pumps and shakes his tail, releasing the fully formed miniature sea dragons. From the moment they hatch, the tiny fry are completely independent of their parents. They have enough yolk to sustain them for a few days, but after that, they must fend for themselves.

Legendary beasts
The leafy sea dragon takes its name from the dragons of mythology. In many cultures dragons are closely connected to the sea; in Chinese mythology, the Dragon King is a godlike figure who controls the oceans.

Dragon antlers are said to resemble those of a stag

> " The appendages of their spines seem to be merely part of the fucus [seaweed] to which they are attached. "
>
> ALBERT GÜNTHER, *An Introduction to the Study of Fishes*, 1880

Up to 300 eggs are incubated in the male's brood patch—a spongy area on the underside of his tail

Developing sea dragon fry are visible through the egg surface

Baby dragons
Sea dragon fry are smaller than a grape when they hatch, tail-first, out of their egg sacs. But they grow fast, reaching 8 in (20 cm) by the time they are one year old.

Eels

Anguilliformes

There are more than 800 species of these elongated, snakelike fish. Some live in fresh water, often venturing out to sea to spawn, but the majority spend their lives in saltwater habitats.

The spiral shape of the Koropepe symbolizes movement and being open to change

Koropepe
This Māori pendant made of whale bone and wax represents a mythical eellike creature called the Koropepe. It represents abundance, prosperity, and new beginnings.

Although they resemble underwater worms or snakes, eels are long, thin fish with gills and one fin. They are usually nocturnal and are found throughout the world in fresh, brackish, or saltwater habitats, including rivers, coastal lagoons, coral reefs, and the open ocean. Eels' thin, extended bodies mean they can fit into small crevices, and they often hide in dark holes, lying in wait for prey, which includes shrimp, crabs, small fish, and invertebrates. When the prey gets close enough, they dart out of their hiding place to snatch it up. They are able to hunt in this way because, unlike most other species of fish, they can swim backward.

Some species make long migrations. European eels can swim more than 6,000 miles (10,000 km) across the Atlantic to the Sargasso Sea to breed. Their larvae drift on the current until young eels reach freshwater bodies in Europe, before becoming ready to start the long migration back across the ocean as adults.

Heads and tails

Eels are important to the Māori people of New Zealand, where they are known as *tuna* and traditionally served as a key food source. In one legend, Tuna, a giant eel who lived in the sky, decided to move to a river. When he arrived, he came across some children playing and ate them. Their furious father, the demigod Māui, chopped off the eel's head and tail, throwing them into the sea and river, respectively. This, it is said, is why there are eels in both salt and fresh water.

According to Japanese mythology, *unagi hime* are shape-shifting eels that live in deep waters. They are said to be able to change their form to that of a beautiful woman to trick strong men into doing battle with their enemies for them.

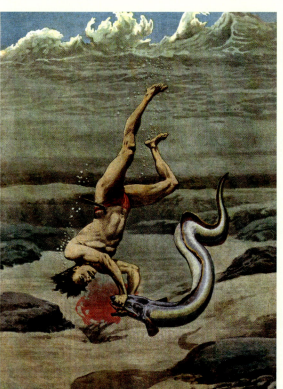

Eel attack
Hawaiian surfer and Olympic swimming champion Duke Kahanamoku (1890–1968) was attacked by an eel while training, as this 1913 illustration depicts.

GIANT CHAIN MORAY EEL
Gymnothorax javanicus
This species has the largest body mass of all the moray eels in the Indo-Pacific ocean, and grows up to 8 ft (2.5 m) long.

RIBBON EEL
Rhinomuraena quaesita
The undulating movement of these colorful eels through the water resembles ribbons in motion.

FINESPOT MORAY
Anarchias leucurus
Found in the Pacific Ocean, the finespot moray—also known as Snyder's moray—is the smallest known species of moray eel.

EELS

Elongated bodies resemble a "garden" of seagrass

Upturned mouth is characteristic of the species

> **Garden eels** secrete a **slime** to strengthen the walls of their burrow

Garden eels
A group of garden eels (Heterocongrinae) poke their heads and bodies out of their burrows to feed on plankton.

Jellyfish

Medusozoa

Jellyfish are gelatinous invertebrates related to corals and anemones. There are thousands of species: the smallest is barely visible to the human eye, while the largest is more than 7 ft (2 m) across and around 120 ft (37 m) long.

Opaque bell can reach more than 3 ft (90 cm) in diameter.

Embossed papier-mâché face of the Gorgon Medusa, who turned anyone who looked at her to stone

Medusa shield, 1897
French zoologist Jean-Baptiste Lamarck, who wrote about jellyfish in *Natural History of Invertebrate Animals* (1815–1822), named them "Méduse" because of their resemblance to the snake-haired Gorgon Medusa from Greek mythology.

Jellyfish are about 95 percent water, have no bones, no brains, no heart, no teeth, and no blood; they are nevertheless complex organisms. They have a highly sophisticated network of cells, called the nerve net, which allows them to detect light, temperature, and chemical changes in the water around them. Some species even have advanced visual systems. Box jellyfish have 24 complex eyes positioned around the base of the bell, some of which have a lens, cornea, iris, and a retina. They can use these to look above the water to help find their way around the tangled mangrove habitats in which they live. Recent studies suggest some jellyfish get tired and sleep at night, and some engage in courtship behavior.

Despite their gelatinous bodies, jellyfish do not always drift passively with the currents. By pulsing their bells, they can swim around, albeit slowly in the case of most species, under their own propulsion. The fastest swimmers are box jellyfish, which are active hunters and can reach maximum speeds of up to 6 ft (2 m) per second.

Lighting the way

Many jellyfish can generate their own light, a phenomenon known as bioluminescence (see pp.286–287). This gives them an ethereal glow in complete darkness. Roman physician and naturalist Pliny the Elder (23–79 CE) described this when he wrote "a walking stick rubbed with the pulmo marinus [jellyfish] will light the way like a torch."

> "The itch they cause has a biting power, just like that of a land nettle."
>
> PLINY THE ELDER, *Natural History*, c. 77 CE

Frilly oral arms can grow to nearly 20 ft (6 m), while the stinging tentacles can reach 25 ft (7.6 m)

Jellyfish have existed for more than 600 million years

Mysterious giant
Despite its huge size, the black sea nettle (*Chrysaora achlyos*) is an elusive species. It occasionally appears in large numbers along North America's Pacific coast.

The function of their bioluminescence is thought to be primarily a way of startling or confusing predators.

Circle of life
Jellyfish have a complex and unusual two-phase life cycle. They start their lives as free-swimming larvae (planulae), which attach to solid surfaces where they transform into flowerlike polyps. As the polyps mature, they bud off into many bell-shaped jellyfish (medusae), which can swim and reproduce. As adults, most jellyfish live for a few months, but some species can live for a year or two. However, one species is much longer lived—the immortal jellyfish (*Turritopsis dohrnii*) is believed to be able to survive indefinitely; it can return to polyp form and repeat its budding process.

Packing a punch
Jellyfish are well known for their ability to sting. Ancient Greek philosopher Aristotle named them *cnidae*, Greek for "stinging." Their trailing tentacles are lined with thousands of stinging capsules known as nematocysts. Each capsule contains a microscopic, needlelike stinger loaded with venom. When a tentacle

FRIED EGG JELLYFISH
Cotylorhiza tuberculata
Measuring roughly 13 in (33 cm) across and named for its resemblance to a fried egg, this species feeds mainly on zooplankton and other jellyfish.

AUSTRALIAN SPOTTED JELLYFISH
Phyllorhiza punctata
Nicknamed the "floating bell" and native between Thailand and Australia, this species feeds on zooplankton, fish eggs, and small fish.

UPSIDE-DOWN JELLYFISH
Cassiopea spp.
This group of species, with a maximum diameter of 12 in (30 cm), typically lie upside down on the seafloor, with "arms" pointing upward.

Origin of a name
The central image in this 1874 engraving by renowned zoologist and artist Ernst Haeckel depicts *Desmonema annasethe*, which Haeckel named after his late wife Anna because its tentacles reminded him of her long hair.

Stinging nettles
The Pacific sea nettle (*Chrysaora fuscescens*) is the most common large jellyfish found off the US Pacific Northwest coast. It can appear in huge numbers, capturing plankton with its long tentacles.

comes into contact with a prey item, the nematocysts fire off within a millionth of a second, injecting the venom into the victim. Once the prey is immobilized, the jellyfish's oral arms transport it to the mouth and on to the stomach. Jellyfish capture plankton, small fish, and other jellyfish in this way. They do not intentionally attack humans, but if a person comes into contact with their tentacles, they will automatically sting. While most species are not harmful to humans, some stings are painful or even fatal. A sting from an Australian box jellyfish (*Chironex fleckeri*) can cause paralysis, cardiac arrest, and death within minutes.

In the ecosystem

Jellyfish play a valuable role within marine ecosystems. They form a critical part of the diet of many predators, including ocean sunfish (*Mola mola*) and leatherback turtles (see pp.88–91), which feed on little else. They provide a habitat within their tentacles for larval fish, and recent research has also revealed their importance in transferring carbon to the ocean floor. However, in recent years, jellyfish numbers have risen and today have become a symbol of ecosystem collapse. Warming seas and pollution are a threat for most animal species, but allow jellyfish to prosper. By removing predatory fish, overfishing also causes jellyfish numbers to soar unchecked.

In bloom

During jellyfish outbreaks, known as blooms, many jellyfish can be found in a single cubic yard, turning the water into a thick soup. Blooms can cause major problems. Many jellyfish eat fish eggs and larval fish, so blooms can decimate fish populations. A single Australian spotted jellyfish (*Phyllorhiza punctata*) can eat up to 2,000 fish eggs per day. The jellyfish also compete with fish and other species in the area for food. Additionally, large quantities of jellyfish can wreak havoc with machinery, clogging boat motors and seawater intake pipes.

RHIZOSTOME JELLYFISH

Upside-down jellyfish (Cassiopeidae) are among more than 70 types of jellyfish that make up the order Rhizostomeae. Unlike other jellyfish, rhizostome species do not have true tentacles or a central mouth. Instead, they have eight branching oral arms, each one adorned with stinging capsules and numerous mouths. The mouths collect food particles and pass them to the stomach.

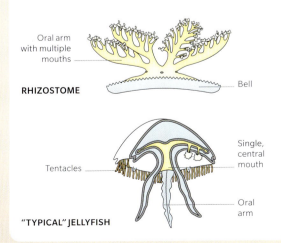

RHIZOSTOME — Oral arm with multiple mouths, Bell
"TYPICAL" JELLYFISH — Tentacles, Single, central mouth, Oral arm

Imprint turns to rock when buried in soft sediment

Trapped in time
Jellyfish rarely fossilize because they have no hard body parts. Scientists believe this fossil from the Ediacaran period (539–635 million years ago) could be a primitive jellyfish.

Dolphins

Delphinidae

Dolphins are known as one of the ocean's most intelligent animals and are famous for their playful antics and "smiling" face. They have been portrayed as both tricksters and saviors in folklore.

> There are **42 known species of dolphin**

Dolphins are marine mammals that can be found in tropical and temperate waters around the world. While most species live in salt water, some have adapted to freshwater environments. Species vary in size from Heaviside's dolphin (*Cephalorhynchus heavisidii*), which measures around 3 ft 3 in (1 m), to orcas (*Orcinus orca*, see pp.222–223), the largest at up to 23 ft (7 m). Their color varies from white and gray to pink, blue, and black. Dolphins, like whales and porpoises, are related to hoofed animals such as hippos. Their ancestors lived on land but moved back into the water around 50 million years ago. As mammals, dolphins have to come to the surface to breathe. They consciously take each breath, even when at rest, which means that only one half of their brain

Ocean acrobats
A group of spinner dolphins leaps out of the water. Their aerial acrobatics may help shake off parasitic fish or allow them to communicate with each other.

DOLPHINS

> "The swiftest ... is the dolphin; it is swifter than a bird and darts faster than a javelin."
>
> PLINY THE ELDER, *Natural History*, c. 77 CE

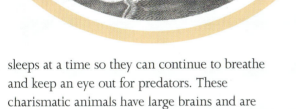

The Monkey and the Dolphin
This 1867 engraving by Gustave Doré depicts Jean de La Fontaine's fable of a dolphin who saves a monkey from a shipwreck before realizing the monkey is lying.

BOTTLENOSE DOLPHIN
Tursiops truncatus

Bottlenose dolphins are perhaps the most archetypal species of dolphin. They have a dark gray back, paler gray belly, and wide mouth, which can look like they are smiling.

sleeps at a time so they can continue to breathe and keep an eye out for predators. These charismatic animals have large brains and are highly intelligent. They can recognize themselves in a mirror, use names for each other, learn tricks, and teach things to other members of their pod.

Dolphins have even learned how to use tools: they have been recorded chasing fish into shells, to trap them, then carrying the shell up to the surface and shaking out their meal. Some have also learned how to place sea sponges over their rostrums (noses) so they do not get scraped by the rocks or coral when foraging on the seafloor.

Tales and traditions

Dolphins are revered in Hawaiian culture because of their association with Kanaloa, an ocean god who sometimes appeared in dolphin form. In Greek legend, the famous musician Arion was captured by pirates, robbed, and forced to throw himself into the sea. Before jumping overboard, he played one last song on his lyre. A dolphin overheard his music and carried him to shore to save him from a watery grave.

In the creation story of the Chumash people from California, Hutash (Mother Earth) created a rainbow bridge to help people cross to the mainland. Some looked down and became dizzy, falling off the bridge, so Hutash turned them into dolphins to stop them from drowning. The Chumash still see dolphins as protectors.

The idea of dolphins as defenders of humans may have some basis in fact. Dolphins have been known to help people in real life; in 2002, dolphins reportedly saved a fisherman in Australia from sharks after his boat sank.

COMMON DOLPHIN
Delphinus delphis

Common dolphins are easily identified by the golden stripe along their side. They often travel together in pods of 10 to 50 individuals, but can sometimes form a mega-pod of up to 10,000 dolphins.

Black-figure decorations depict mythical scenes

Dionysian dolphins
This Greek hydria (jug), from the early 6th century BCE, shows the god Dionysus turning Tyrrhenian pirates into dolphins after they planned to kidnap him.

SPINNER DOLPHIN
Stenella longirostris

With their long, slender bodies, spinner dolphins are perfectly evolved for acrobatics. They are known for their extraordinary ability to leap out of the water, spinning several times before landing.

Orca

Orcinus orca

Also known as killer whales, orcas are among the ocean's most easily identifiable animals. These intelligent creatures are efficient predators that can kill a great white shark.

Found all around the world oceans, orcas are one of the sea's apex predators. They were given the name "killer whales" by sailors in ancient times because they were known to eat large whales. There is only one species of orca, but there are different variants, called "ecotypes," which specialize in hunting different prey. Some orcas eat fish, such as salmon and mackerel, and squid, while others hunt sea birds and marine mammals, including seals, sea lions, and other whales. Orcas have been seen eating blue whale calves and some are even known to kill great

Brown stripes and patches
instead of the orca's iconic black and white patterns

Orca vessel
This Nazca orca effigy from 100 BCE–600 CE has spots of blood on its teeth and a human head—possibly decapitated—painted on its underside.

Killer whales
Sometimes called the "wolves of the sea," orcas are expert hunters. As well as eating fish, some populations hunt large marine mammals and sharks.

white sharks, ripping out and devouring the shark's nutrient-rich liver and leaving the rest of the carcass.

Orcas often live in matrilineal societies, led by an older female. A mother nurses her calf for around a year but the offspring stays with her into adulthood. They are among the few mammals known to go through menopause (alongside humans, narwhals, pilot whales, and beluga whales). Female orcas can live to around 90 years old, compared to males, which only reach around 60 years. Researchers believe female orcas have evolved to live so long after menopause because the grandmothers' knowledge increases the chances of their children and grandchildren surviving.

Intelligent and highly respected

Within the pod, orcas have complex social structures and even form friendships with each other. They are highly intelligent and have developed ingenious ways of catching their prey, which they teach to their pod mates so the knowledge passes down through the generations. They are known to create huge waves to wash seals off the ice and into the water where they are easier to catch. They have also learned to ride waves into shore to hunt seals resting on the beach, before waiting for another wave to take them back out to sea. Orcas are the only whales known to beach themselves in this way on purpose.

Orcas are deeply respected by many Indigenous nations. In Haida mythology, they are said to be the rulers of the underworld, which included the sea. Some cultures believe that an orca sighting signals the return of an old chief or that a loved one is trying to communicate from the other side. In Kwakwaka'wakw legend, humans and orcas used to hunt each other, until a young boy became friends with the orcas and facilitated a truce so humans and orcas would stop killing each other and live in harmony. The Kwakwaka'wakw people still respect the truce to this day.

Weird and wonderful
First published in a book in 1613, this strange illustration of an orca is by Italian naturalist Ulisse Aldrovandi, who was known for his detailed, but not always realistic, depictions of flora and fauna.

Orcas can **swim** at speeds of **35 mph (56 kph)**

ECHOLOCATION

Like other toothed whales, orcas use echolocation to navigate, communicate, and find food. They create a series of clicks using a part of their nose called "phonic lips." If the sound waves from these clicks hit an object in the water, they bounce back, providing the orca with information about the object's size and location.

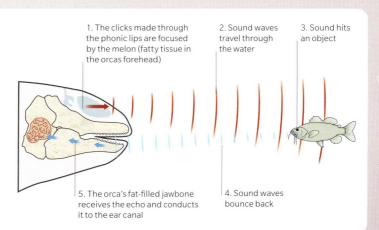

1. The clicks made through the phonic lips are focused by the melon (fatty tissue in the orcas forehead)
2. Sound waves travel through the water
3. Sound hits an object
4. Sound waves bounce back
5. The orca's fat-filled jawbone receives the echo and conducts it to the ear canal

ORCA USING ECHOLOCATION TO FIND FOOD

Natural exfoliation
Sperm whales (*Physeter macrocephalus*) naturally shed their skin so they do not become covered in parasites. These social animals often rub up against each other, possibly to help exfoliation.

> A **blue whale's heartbeat** can be heard by sonar equipment **2 miles (3 km)** away

Whales

Cetacea

There are around 15 species of baleen whale and more than 70 toothed whale species that live around the world. These intelligent species include the largest mammals on Earth.

There are two types of whales: toothed whales (suborder Odontoceti), which include sperm, beluga, narwhal, pilot, and beaked whales, and baleen whales (suborder Mysticeti), such as blue, fin, humpback, gray, minke, and right whales. Toothed whales, as the name suggests, have teeth, which they use to catch and eat their food, usually fish and squid. They find prey using echolocation (see p.223).

Baleen whales feed on a variety of food in the water, such as small fish, krill, and plankton. They have plates called "baleen" made of keratin—the same protein human hair is made of—in their mouths that they use like a giant sieve when feeding. They take a mouthful of water and use the baleen plates to filter out the water while keeping their food in their mouth to eat. Another difference between the two types is that toothed whales have one blowhole on the top of their head, while baleen whales have two.

Whale species vary considerably in size. The blue whale (*Balaenoptera musculus*) is the largest animal ever to have lived—blue whales can measure up to around 100 ft (30 m). Notable smaller species include the dwarf sperm whale (*Kogia sima*), which can measure around 9 ft (2.7 m).

From the poles to the tropics
Whales are found in all regions of the ocean, from narwhals (see pp.178–179) in Arctic seas to common Bryde's and Omura's whales (*Balaenoptera brydei* and *Balaenoptera*

American flag
flies above tents

Scrimshaw whale tooth
Scrimshaw is engraving on bone or ivory. This US military scene has been carved on a whale's tooth.

Breaking the surface
A humpback whale (*Megaptera novaeangliae*) breaches in the Sea of Cortez, Mexico. They may do this for communication, to remove parasites, or for fun.

> " Is it not curious, that so vast a being as the whale should see the world through so small an eye ... ? "
>
> HERMAN MELVILLE, *Moby-Dick*, 1851

LONG-FINNED PILOT WHALE
Globicephala melas
Pilot whales are known for their strong social bonds, which also make them susceptible to mass strandings.

BLUE WHALE
Balaenoptera musculus
At around 108 ft (33 m), blue whales are the largest animals on Earth. Calves are around the same size as an adult African elephant.

COMMON MINKE WHALE
Balaenoptera acutorostrata
This baleen whale grows to around 36 ft (11 m) long and has a distinctive pale stripe on its front fins, making it easily identifiable.

omurai), which prefer tropical waters. They can also travel large distances. Humpback whales can swim 5,000 miles (8,000 km) between their feeding and breeding grounds and gray whales migrate farther than any other mammal, swimming 12,000 miles (19,000 km) from the Arctic to Baja California, Mexico, and back during their annual migration.

As mammals, whales must come to the surface to breathe air, but they can still descend to the ocean depths. The deepest recorded dive of any whale is the Cuvier's beaked whale, which swam 9,816 ft (2,992 m) deep in 2011.

Whale society

Whales are intelligent and social. Many species communicate, hunt, and play together as well as care for each other. They can form friendships that last for several years. Their intelligence and close bonds mean that some whale species also have culture—they learn different behaviors from each other and pass them down to future generations. For example, humpback whales can learn complex songs from each other and sperm whales live in distinct clans that have their own dialects.

Whales communicate through sound. Baleen whales make low-frequency moans, groans, and calls in sequences we know as whale song, while toothed whales make high-pitched clicks and whistles. It was only recently discovered that whales could sing. Although early sailors could hear whales, they did not know what the noises were. In the 1950s, the US Navy was trying to listen for Soviet submarines and accidentally captured the sounds of whales. It was not until 1967, when marine biologist Roger Payne first recorded whale song, that these sounds were shared with the public.

Cultural representations

Whales have appeared in mythology around the world as devilish beasts and sacred spirits. Medieval manuscripts, called "bestiaries," described them as huge sea monsters that dragged unsuspecting sailors to the bottom of the ocean. In Japanese myth, *Bake-kujira* were ghostly whale skeletons that returned to Earth to take revenge by cursing the people who killed them. In 13th-century Norse literature, a sea creature called *hafgufa* lured fish into its wide-open mouth by releasing a special perfume. Researchers later realized this was inspired by trap feeding: whales open their mouths and fish are drawn in, mistakenly thinking they have found shelter.

The 1851 novel *Moby-Dick* by Herman Melville is one of the most famous stories of monstrous whales: a fictional albino sperm whale is pursued across the seas by Captain Ahab after biting off his leg. It is said to have been inspired by real-life sperm-whale attacks on whaling ships in the 1800s.

There are also many positive representations of whales. Inuit peoples honor them for providing food, in the form of meat, and light from their oil. The Māori revere whales as descendants of the ocean god, Tangaroa, and in Hawaiian culture, they can represent the god of the ocean, Kanaloa. In the Bible story, God saved Jonah from drowning by sending a "big fish" (thought to be a whale) to swallow him. He spent three days in its belly before being vomited up.

Yunus and the whale
Similar to the story of Jonah, the Qur'an tells of the prophet Yunus, who was swallowed by a large fish or whale, illustrated here in the 14th-century book *Jami' al-Tawarikh* by Rashid-al-Din Hamadani.

BUBBLE-NET FEEDING

Humpback whales and Bryde's whales use a hunting strategy called bubble-net feeding. One whale swims in circles while blowing bubbles to create a barrier that herds the fish. From below, the others work together to drive the fish into this "net" to trap and eat them.

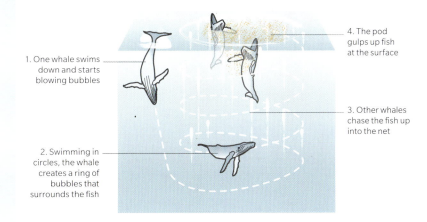

1. One whale swims down and starts blowing bubbles
2. Swimming in circles, the whale creates a ring of bubbles that surrounds the fish
3. Other whales chase the fish up into the net
4. The pod gulps up fish at the surface

From the depths to the air
A huge humpback whale (*Megaptera novaeangliae*) takes a leap out of the water. Scientists are unsure why these and other cetaceans—including sperm whales (*Physeter macrocephalus*)—exhibit this breaching behavior. It is thought it may help them communicate, stun prey, or dislodge parasites. Humpbacks are also known for their haunting whale song—complex patterns of repeated sounds, used for a variety of purposes, including communication, navigation, and probably courtship.

Sea creatures surround the two Titans

Ocean deities A mosaic from the ancient city of Zeugma in Türkiye shows the Titan Oceanus, god of the great sea, and his wife, Tethys.

Atlantic Ocean

The zig-zag shaped Atlantic is the second-largest and second-deepest ocean basin in the world. Covering around one-fifth of the Earth's surface, it stretches from the equator to the icy polar regions.

The Atlantic lies between Europe and Africa to the east and the Americas to the west. With its dependent seas, it covers an area of around 32,870,000 sq miles (85,133,000 sq km), and its deepest point is 27,585 ft (8,408 m) in the Puerto Rico Trench. It takes its name from the Atlas Mountains where, according to Greek mythology, the Titan Atlas stood as he held up the heavens. To the ancient Greeks, these mountains marked the western edge of the world, beyond which lay the great, endless river surrounding the Earth. Another Titan, Oceanus, ruled this realm, and so became associated with the Atlantic.

Although Christopher Columbus is famous for his Atlantic explorations, beginning in 1492, no one knows who was first to cross it. In the 6th century, Irish monk Brendan the Navigator is said to have found unknown lands while sailing west from Ireland. Around the year 1000 CE, Leif Eriksson is known to have made landfall in what is now Canada. Crossing the Atlantic by sail is made possible by the circular pattern of currents and winds that are often called the trade winds. The Coriolis effect (see p.23), caused by the spinning of the planet, creates westward winds near the equator, which then swirl around and back to the east as they move toward the poles. For centuries after Columbus, European invaders and traders used these routes—most notoriously for shipping enslaved Africans to the Americas, and carrying looted treasure, or cargoes of sugar and cotton, back to Europe.

Atlantic Storm, **1876**
This painting by John Singer Sargent (1856–1925) evokes the fear of the Atlantic's powerful waves, experienced by Sargent on a crossing from New York to Liverpool.

BENEATH THE WAVES

Agwe and the mermaid
Haitian water spirit La Sirene is shown with her husband, Met Agwe, in this work by Haitian painter André Pierre (1914–2005). La Sirene is said to live in a treasure-filled undersea palace.

MID-ATLANTIC RIDGE

The Mid-Atlantic Ridge is a volcanic spreading ridge that separates the North American and Eurasian tectonic plates in the north, and the South American and African plates in the south. As magma pushes up and cools, it creates new seabed, widening the ocean by 0.8–2 in (2–5 cm) per year. When the ridge breaks the surface, it forms islands.

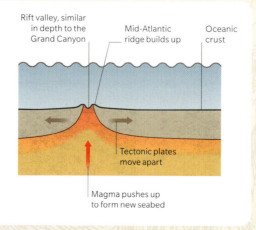

CROSS SECTION OF THE MID-ATLANTIC RIDGE

200 million years ago, **South America** was joined to **Africa**. Their **coastlines** still match

Although it has now been widely explored, the Atlantic still poses many dangers, from hazardous icebergs to the towering waves of the Bay of Biscay, and the powerful hurricanes that swirl westward across the tropics every summer.

Although not as island-filled as the Pacific, the main part of the Atlantic has a number of key island groups and chains. As well as Iceland, islands along the mid-Atlantic ridge include the Azores, Ascension Island, Saint Helena (famous as the place where Napoleon spent his final years), and the even more remote Tristan da Cunha. The Atlantic's larger islands include Newfoundland, Great Britain, and Ireland, while smaller volcanic islands such as Bermuda, the Cape Verde islands, and the Canaries were important staging posts for shipping during the age of sail.

In and around the Caribbean, from the 1500s to the 1700s, some islands became pirate hideouts. Independent pirate captains such as Blackbeard (real name Edward Teach), and corsairs, who raided ships on behalf of their governments, preyed on the many cargo ships carrying silver and gold back from colonized lands in the Americas. They would also steal weapons and gunpowder, tobacco, or enslaved people, as well as taking over the ships themselves.

Shared stories

Traditional foods, folklore, and fashions of the Caribbean islands are often similar to those of West Africa, due to the movement of enslaved people. One example is the African water goddess or spirit Yemoja, often known as Mami Wata, who has the power to bring wealth and success, or sometimes disaster. In Haiti in the Caribbean, her equivalent is the exuberant sea goddess La Sirene. Both goddesses are often shown in carvings and paintings with mermaid tails, and are said to live beneath the sea. According to folklore, they may sometimes abduct a human, who returns from the sea with magical powers or great wisdom.

Snakes symbolize Mami Wata's ability to charm snakes

Mami Wata
This Nigerian Yoruba wooden carving of water goddess Mami Wata shows her with a fish tail and two snakes.

> " The rock that is in the middle of the sea does not fear the rain. "
>
> ZAMBIAN PROVERB

Dramatic landscape
The Faroe Islands' largest lake, Sørvágsvatn, lies 130 ft (40 m) above the Atlantic, flowing into the ocean via a waterfall.

BENEATH THE WAVES

Echinoderms

Echinodermata

This group of marine animals includes sea cucumbers, starfish, brittle stars, feather stars, and sea urchins. The name means "spiny skin" or "hedgehog skin" in Greek.

Fan and feathers
This bright red Gorgonian fan coral in waters near Papua New Guinea is covered in feather stars.

ECHINODERMS

Echinoderms are marine invertebrates that live on the seafloor in both shallow and deep seas worldwide. These ancient animals evolved before the dinosaurs at least 540 million years ago and, so far, around 13,000 different fossil species of them have been discovered.

There are around 7,000 species of echinoderm in existence today. They are separated into 21 different classes, based mainly on the structure of their skeleton. All echinoderms have five-fold radial symmetry, spiny skin, and nerve nets (a set of interconnected neurons and a nervous system) instead of a centralized brain. The best-known are starfish—also called sea stars—which, like all echinoderms, have the power to regrow lost limbs and some organs, as long as some of the severed part is still attached to its central disk.

Echinoderms are generally small, beginning at around 4 in (10 cm) in length or diameter. Some, like the tiger tail sea cucumber (*Holothuria thomasi*), however, can grow to 6½ ft (2 m) in size. The stem of some extinct sea lily species grew to 66 ft (22 m) in length.

Starfish are known for their slow movement: the speediest species—the sunflower sea star (*Pycnopodia helianthoides*)—can cover 3 ft (1 m) in one minute. Sea urchins cannot swim, and so they use their spines and appendages—called "tube feet"—to move around on the seabed (see pp.66–67).

Ocean gardeners

Echinoderms play an important role in marine habitats. Grazing species such as sea urchins keep algae in check and prevent them from smothering coral reefs. Sea cucumbers clean the water by taking in organic matter and excreting nutrient-rich sand, which recycles vital materials and keeps the whole ecosystem healthy. However, species can become too numerous: the crown-of-thorns starfish (*Acanthaster planci*) has disrupted many coral reefs—including the Great Barrier Reef (see pp.210–211)—by feeding on their living polyps, the marine invertebrates that make up coral.

Humans have eaten echinoderms—mainly sea urchins—throughout history. There are records of the ancient Greeks and Romans eating them, and many cultures still consume them today. Around 80 to 90 percent of the global demand for sea urchins comes from Japan. In China, cooked and dried sea cucumbers are popular. Usually eaten in soups, they are known as *haishen* or *trepang*.

Stars on the seabed
This 1887 engraving by French artist Alphonse Demarle depicts echinoderms including sea lilies, feather stars, and starfish.

SEA APPLE
Pseudocolochirus violaceus
These sea cucumbers, named for their likeness to apples, are found in Indo-Pacific waters.

BANDED SEA URCHIN
Echinothrix calamaris
Also known as the double spined urchin, this urchin's spines give a sting like a bee's.

FEATHER STAR
Crinoidea
Feather stars can swim through the water by waving their arms up and down.

Brittle, dried-out sea cucumbers have a water content of just 8–10 percent

Food source
Dried sea cucumber is a delicacy in many Asian countries, where it is also used in traditional medicine.

Some **deep water sea cucumber** species can **swim**

Rocky Reefs

Rocky reefs are coastal outcrops of rock that are either periodically or permanently submerged by the ocean. They form the basis of varied ecosystems for algae, plankton, fish, birds, and more.

Warning lights
Lighthouses such as this one at Tourlitis, off the island of Andros, Greece, are often built on rocky reefs, allowing them to project out into the water.

Rocky reefs are found worldwide, from tropical to polar regions. Sometimes referred to as geogenic ("rock-derived") reefs, they develop in shallow water on shorelines where cliff faces and rocky shore platforms are weathered and broken up by wind and wave action.

Different microhabitats develop on rocky reefs, depending on the local climate and prevailing water levels. Rocky reefs in the upper shore that are exposed by changing tides, for example, will typically host organisms that can survive prolonged exposure to air and sunlight, such as limpets and sea snails. Seaweeds and small fish such as blennies thrive lower down the shore, where there is more water cover, while crabs and anemones will shelter in rock pools (see pp.64–65). In constantly submerged areas, fish species may include wrasse (Labridae) and wolf eels (*Anarrhichthys ocellatus*), and kelp and other seaweeds.

Reef stars
A common sun star (*Crossaster papposus*) and common brittle stars (*Ophiothrix fragilis*), lie on a reef off the Shetland Islands, UK.

> " So we sailed on through the narrow straits, crying aloud for fear of Scylla on the one hand while divine Charybdis sucked the sea in terribly on the other. "
>
> HOMER, *The Odyssey*, c. 725–675 BCE

Rocky lifestyle

The rocks provide a stable surface for algae and for sessile (immobile) animals such as barnacles, sponges, and cold-water corals. The algae, in turn, are food for sea urchins and other species. Currents bring in plankton and larvae, and the rock crevices provide shelter that fish and other marine animals use for breeding and nursery areas. As a result, these habitats form rich and ever changing hunting grounds for shore birds and marine mammals. Humans, too, use rocky reefs to harvest seaweed, crustaceans, and fish, despite them being often hazardous environments in which to swim and sail. More recently, they have become sites for leisure pursuits such as diving. In some places, however, such as the Firth of Lorn in Scotland and Cape Arago in Oregon, human and economic activity on or around them is managed due to the reef's scientific and ecological importance.

Scylla and Charybdis
This 3rd-century BCE flask depicts the sea monster Scylla. In Greek legend, Scylla and the whirlpool Charybdis occupied the submerged rocks in the Strait of Messina, in Italy.

Scylla had a female torso and six long necks, each one ending in a head with three rows of teeth

Deep-sea deity
Divers swim past a statue of the god Dionysus as they explore the submerged Roman town of Baiae, off the Italian coast.

Sunken ruins
Captain Nemo and Professor Aronnax gaze upon the remains of Atlantis in this illustration from Jules Verne's 1869 novel *Vingt Mille Lieues sous les mers* (*Twenty Thousand Leagues Under the Sea*).

Sunken Cities

The idea of cities that lie below the ocean is an enduring theme in legend and fiction. However, there is some truth behind the notion: in antiquity, several cities became submerged through natural disasters.

Egyptian goddess of music, Hathor

For two millennia after Greek philosopher Plato wrote about the "lost" land of Atlantis in the 4th century BCE in his books *Timaeus* and *Critias*, it was thought he drew on ancient tales and folk memories concerning a place that actually existed. It is now believed his story of an island-city swallowed by the ocean for its inhabitants' failure to honor the gods was a morality tale; efforts to locate it—off the Greek island of Santorini, or the Azores, Mauritania, and the North Pole—have all failed.

In the centuries after Plato described Atlantis, the entire Egyptian port city of Thonis-Heracleion collapsed into the Mediterranean following several natural disasters over a few hundred years. It was not rediscovered until 2000. Similarly, the Roman spa town of Baiae began to sink into the Bay of Naples. This was caused by bradyseism, a geological process whereby volcanic activity at nearby Vesuvius caused the land around Baiae to rise and fall. By the 8th century CE, the lower part of Baiae was submerged. Other sunken settlements include the Neolithic village of Atlit Yam in Israel, which was lost to rising sea levels more than 8,000 years ago. The 5,000-year-old community of Pavlopetri, off Laconia in Greece, became submerged for reasons that are still unknown.

Underwater living

Today, sunken settlements are deliberately being built. The Aquarius Reef Base sits 60 ft (18 m) below the waves off the coast of Florida and is the world's only undersea laboratory. Fabien Cousteau—grandson of oceanographer Jacques Cousteau—plans to build an underwater research facility called Proteus in the Caribbean, where scientists and visitors will be able to live and study marine ecosystems.

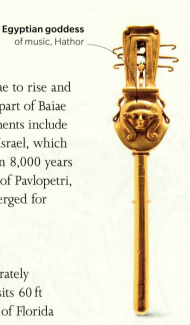

Buried treasure
This gold artifact is a sistrum, a rattle-like instrument found at the sunken city of Thonis-Heracleion, off Alexandria, Egypt, in the Mediterranean.

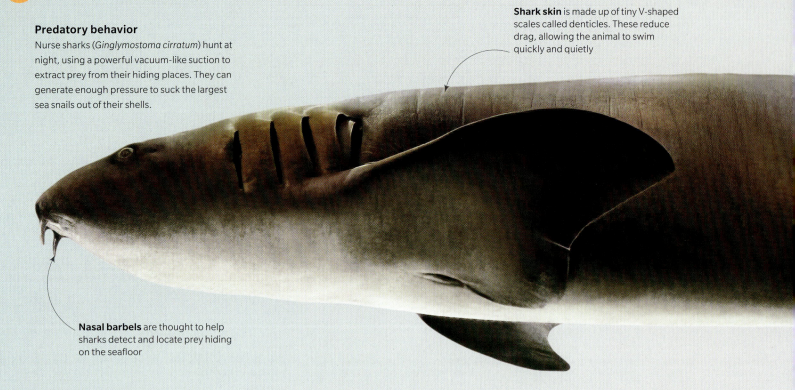

Predatory behavior
Nurse sharks (Ginglymostoma cirratum) hunt at night, using a powerful vacuum-like suction to extract prey from their hiding places. They can generate enough pressure to suck the largest sea snails out of their shells.

Shark skin is made up of tiny V-shaped scales called denticles. These reduce drag, allowing the animal to swim quickly and quietly

Nasal barbels are thought to help sharks detect and locate prey hiding on the seafloor

Sharks

Elasmobranchii

Although sharks are best known as fearsome, hulking predators, these ancient marine animals are varied in both size and feeding habits.

Fossil records show that sharks have been on Earth for at least 450 million years. They outlived the dinosaurs and survived five major extinction events. Today, more than 500 species of shark inhabit the ocean waters, ranging from a couple inches—the dwarf lantern shark (*Etmopterus perryi*)—to the 60-ft- (18-m-) long whale shark (*Rhincodon typus*). Around half of all shark species grow to no more than 3 ft (1 m) in length. Sharks are found in a range of habitats worldwide—from the depths of the ocean to shallow tropical coral reefs, sandy plains, and under Arctic ice. Cold-water species such as salmon sharks (*Lamna ditropis*) can raise their body temperature above that of the water by using a network of blood vessels that retain the heat produced by their muscles. Some sharks are fast-moving predators and can swim through the water at speeds of up to 45 mph (72 kph) to catch their prey, while others are filter feeders that swim slowly and survive on tiny plankton. Great white sharks (*Carcharodon carcharias*) are known for high-speed attacks on seals; other shark species feed on fish, mollusks, and crustaceans. Some sharks also feed on smaller sharks, which means that populations may segregate by size.

Flexible frame
Despite their differences, all sharks share some common traits, including the fact that they do not have bones. Instead, their skeletons are made

Elongated features give fierce appearance

Shark god
This carving of one of the two Fijian ancestral shark gods dates from the 18th century. His name, Dakuwaqa, means "back of the boat."

SHARKS 241

Upper lobe of the tail fin is significantly longer than the bottom lobe, measuring approximately one quarter of the nurse shark's entire body length

" ... this animal is just as frightened at man, as man is at it ... "

PLINY THE ELDER, *Natural History*, (77 CE)

Tip carved from bone is attached to wooden core

Fishing hook
In Hawai'ian cultural tradition, fishhooks—known as *makau*—have long been symbols of good fortune and strength. The hooks were fashioned to catch many different forms of sea life, including sharks.

of cartilage. It is thought that cartilaginous skeletons evolved because they are lighter and more flexible, giving sharks the ability to swim faster and more efficiently. The teeth are the only part of a shark skeleton that is not made from cartilage.

Hunting senses
Sharks have a powerful set of senses for locating and capturing their prey. Hearing and smell are most important for detecting prey at a distance. A shark's ear is similar to a human's but much more sensitive, and most sharks have a powerful sense of smell. They are able to pick up minute chemical traces in the water and even tell which direction they are coming from. Some studies have reported that blacktip sharks (*Carcharhinus limbatus*) can detect fish flesh in concentrations of one part per 10 billion parts of seawater.

As a shark closes in on its prey, it starts to make greater use of its eyes, which have a layer of mirrored crystals behind their retinas, known as the tapetum lucidum, that allows them to see in lowlight, murky waters. In the final approaches, its lateral line system comes into play. This is a series of fluid-filled tubes that run along the length of a shark's body, in which cells sense the vibrations and pressure changes in the water made by any moving target.

BENEATH THE WAVES

BASKING SHARK
Cetorhinus maximus

The second-largest ocean fish, basking sharks can grow to 40 ft (12 m). They feed on tiny plankton, filtering up to 2,000 tons (1,800 metric tons) of seawater per hour.

ZEBRA SHARK
Stegostoma tigrinum

These nocturnal sharks feed on mollusks, crustaceans, fish, and even sea snakes in the tropical Indo-Pacific. The young are striped, but adults are spotted.

GOBLIN SHARK
Mitsukurina owstoni

An elusive, deep-water shark that grows to more than 13 ft (4 m), the goblin shark's jaws can be thrust forward at great speed to capture prey.

SHARKS

Bone crusher
The horn shark (*Heterodontus francisci*) is a small, slow species, but has one of the most powerful bite forces, relative to size, of any shark. It uses this to crush its prey—mollusks, echinoderms, and crustaceans.

Dorsal fins contain sharp spines, which help deter predators such as larger sharks and fish

Pectoral fins are used to crawl along the ocean floor

Ocean giants
Whale sharks are the largest fish on the planet, but they feed only on the oceans' smallest animals—tiny plankton that they filter from the water.

Sharks (along with their close relatives, skates and rays) also have the ability to pick up the faint electrical fields that animals produce when they move around. They detect these signals using jelly-filled sense organs on their heads known as ampullae of Lorenzini, which are modified parts of the lateral line system.

Perception and reality
Sharks are often portrayed in fiction and media as deadly killers that will unthinkingly attack humans. Notably, director Steven Spielberg's infamous 1975 film *Jaws*, a thriller about an enormous great white shark that kills a number of swimmers, sparked a wave of terror among the public on its release. Pioneering US marine biologist Eugenie Clark (1922–2015), known as "The Shark Lady," was among the first to promote a more nuanced understanding of sharks. Through her work and public engagement, she helped dispel myths and show that most species represent no danger to humans.

Sharks and the environment
Not only do sharks pose little threat to people, but they are central to the health of the world's oceans and climate. As apex predators, they keep prey animal populations in check, and maintain the balance of the environment. The loss of sharks can have catastrophic impacts on the ecosystems in which they live: shark populations around the world are shrinking as a result of trophy hunting, the demand for shark fin soup, and being unintentionally caught as bycatch by fishing boats. The climate crisis has the potential to deepen the problem even further, as warming waters become less habitable and cause the range of various shark species to shrink.

SHARK TEETH
Sharks can have up to 15 rows of teeth, although only the front two rows are functional. The others are folded back against the inside of the jaw until they are needed. The teeth do not have roots, but are attached by connective tissue. Over time, the tissue moves forward, bringing new teeth to replace missing ones. A single shark can shed 50,000 teeth during its lifetime.

CROSS SECTION OF A SHARK JAW

Front row — Second row — Soft tissue connects tooth to jaw — Lower jaw

Stingrays

Myliobatoidei

Related to sharks, stingrays are cartilaginous fish with round, flat bodies and a long tail with a venomous spine. When under threat, they use this in self-defense to deliver a sting that can be fatal.

Stingray City
A southern stingray (*Hypanus americanus*) visits Stingray City in Grand Cayman. This tourist spot is known for the many stingrays that congregate in the hope of being fed.

Stingrays have been on the planet for **150 million years**

Mouth on underside of body makes it easier to eat bottom-dwelling prey

Found in tropical and subtropical waters worldwide, stingrays have a round body with their eyes on the top and mouth underneath. They usually swim along the seafloor and move through the water by undulating their fins to propel themselves forward. When they find prey, including fish, crustaceans, and mollusks, they trap it under their bodies and create suction to draw the food to their mouth.

Usually timid, all stingrays—but not all rays—have venomous spines at the end of their tails, which they use to protect themselves from threats. In areas where stingrays are common, people often walk through shallow water using the "stingray shuffle," shuffling the feet through the seabed rather than taking steps, to avoid getting accidentally stung by a ray that might be hiding in the sand.

Venomous spines

The largest known marine stingray is the smalleye stingray (*Megatrygon microps*). It can grow to 8 ft 3 in (2.5 m) wide, but it has tiny eyes no bigger than raisins. It has a pattern of white dots scattered across its back that can be used to identify each individual. Although little is known about the smalleye stingray because they are rarely seen by humans, researchers at the Marine Megafauna Foundation witnessed its self-defense mechanism: when trying to tag one, they discovered it could bring its barb over its head and swing it around, like a scorpion, aiming for the threat.

Stingrays and their deadly barbs also feature in many legends. In Greek mythology, a prophecy told that death would come to the hero Odysseus from the sea. This came true when his son Telegonus did not recognize him and, during a fight, fatally injured him with a spear tipped with a stingray's barb. There are differing accounts of how the Greek hero Hercules lost his finger: according to some stories, it was bitten off by a lion, while others state it was the swipe of a stingray's barb. In Māori folklore, Whaitere, a magical stingray, protected the ocean from overfishing after her parents were both killed by human fishers.

Kite surfer
The common stingray (*Dasyatis pastinaca*) has a kite-shaped body, as illustrated in this 19th-century print.

BLUESPOTTED RIBBONTAIL RAY
Taeniura lymma
The bright blue dots covering this ray's body provide a warning to potential predators.

LEOPARD WHIPRAY
Himantura leoparda
This ray is recognizable for its striking pattern and impressive tail, which can be three times as long as its body.

RED STINGRAY
Hemitrygon akajei
This ray was named for its deep reddish-brown color, which can be orange near the fins.

Rays beneath the waves
Beating their wings just beneath the waves, a group (known as a "fever") of hundreds of Munk's devil rays (*Mobula munkiana*) swim through Magdalena Bay, off the Mexican Pacific coast. Members of this and other species of stingrays have been known to jump out of the water, with rays leaping and somersaulting both alone and in groups above the surface.

Primordial seas
This 1940s lithograph imagines what the seabed might have looked like in prehistoric times. Bottom left is a trilobite, an extinct marine arthropod.

Ocean Floor

The ocean floor has geographical features in common with the land above water, but only around 25 percent of the world's seas have been mapped, or even seen, by humans.

The seafloor is an inaccessible place to map, especially at great depths. Much of what is known about the floor of the ocean has been learned through bathymetry—the measurement of the depth of water in the ocean, lakes, and rivers. These measurements were historically made by throwing a rope over the side of a vessel and measuring the length needed to reach the bottom, a cumbersome and inaccurate process. Today, multi-beam sonar arrays fitted underneath ships send out multiple sound waves in a fan-shaped pattern, building up a picture of the seabed. In addition, the ocean surface may look flat to us but it has bumps and dips that reflect those on the ocean floor, and radar altimeters on satellites can measure these.

This technology has revealed that the ocean floor is a varied landscape of mountains, volcanoes, waterfalls, and trenches (see pp.300–301). The world's tallest mountain is thought to be Mauna Kea on the Island of Hawai'i, which is 33,500 ft (10,210 m) tall, even though only 13,803 ft (4,207 m) of it is above sea level. The largest waterfall in the world, the Denmark Strait cataract, is also found on the sea bed (see p.284).

There are some phenomena that only exist on the ocean floor. These include deep-sea hydrothermal vents (see pp.294–295) and methane seeps—microbially rich areas where subsurface gas escapes from cracks in the volcanic rock that makes up most of the ocean bed.

There are more than **1 million volcanoes** on the ocean floor

The ocean floor is mostly covered with sediment, with an average thickness of 1,500 ft (450 m). Studies of undisturbed sediment via drilling and core samples have uncovered information on the Earth's past climate, magnetic field, and tectonic movements.

New life, new materials
Researchers also use remotely operated robots, gliders, submersibles, and sensors to study seafloor habitats and the species that live in them. In 2024, researchers from the Schmidt Ocean Institute discovered more than 100 species that are probably new to science during a month-long expedition in deep waters off the coast of Chile and Peru. These included corals, sponges, sea urchins, mollusks, and crustaceans. Metals and minerals are contained in the seafloor, including copper, zinc, nickel, gold, silver, and phosphorus. Some companies are assessing the possibility of mining seabeds, while ecological groups are lobbying against this.

Human debris
Fish swim over the *Virgo* shipwreck on the sea bed near Recife, Brazil. Ships, planes, and military objects are common wrecks on the ocean floor.

Eyes move to the same side of the head between larval stage and adulthood

Sea bed camouflage
Flounders (Platichthys) have evolved sediment-colored skin to hide from predators on the sea bed.

Twilight Zone

Also known as the mesopelagic zone, this part of the ocean is a world of near-complete darkness, where the sun's energy struggles to penetrate, food is scarce, and animals must rely on scraps or daily migrations to survive. Eighty percent of the fauna here generates its own light via bioluminescence.

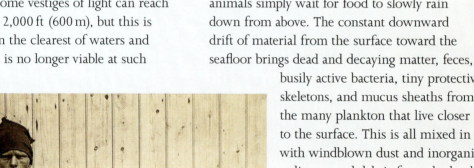

Branching arms have tiny sharp hooks to help it capture prey—mainly zooplankton that get stuck in its tangled limbs

Only one percent of the sun's light energy reaches the top of the twilight zone, which starts 660–850 ft (200–260 m) below the surface of the ocean. Some vestiges of light can reach as far down as 2,000 ft (600 m), but this is only possible in the clearest of waters and photosynthesis is no longer viable at such depths. At the end of the twilight zone, at 3,300 ft (1,000 m), no light remains. In this cold world, so alien to humans, many animals simply wait for food to slowly rain down from above. The constant downward drift of material from the surface toward the seafloor brings dead and decaying matter, feces, busily active bacteria, tiny protective skeletons, and mucus sheaths from the many plankton that live closer to the surface. This is all mixed in with windblown dust and inorganic sediment and debris from the land.

Many of the inhabitants of the twilight zone migrate upward every day to feed at the surface in the relative safety of night, retreating back down to the depths at dawn in an effort to avoid predation. These migrants include some species of copepod—minuscule crustaceans only $^{3}/_{64}$–$^{5}/_{64}$ in (1–2mm) in length—who travel the proportionately huge distance of several hundreds of yards each day. They also face the threat of being eaten by large numbers of lantern fish (Myctophidae), who closely track them as they swim. The most abundant fish at these depths is the bristlemouth

Linking arms
Many species of basket star—a type of echinoderm with multiple branching arms that form a basketlike network—inhabit the depths, including this *Gorgonocephalus* from the Indian Ocean.

Deep-sea pioneer
Gordon Leslie Challen (1887–1955), a master shipwright and diver from Port Stanley, Falkland Islands, stands beside his diving suit, helmets, and air pump.

Yellow bioluminescence in its short arms is used to signal to a potential mate

> " In the ocean, bioluminescence is the rule rather than the exception. "
>
> EDITH WIDDER
> *Below the Edge of Darkness,* 2021

Twilight dweller
The small, bioluminescent diaphanous pelagic octopod (*Japetella diaphana*) lives its entire life in the twilight zone, never touching the seafloor and mating near the base of the zone.

(Gonostomatidae). These small animals, about the size of a human's little finger, are bioluminescent and have a large mouth full of fangs. They are thought to be the most numerous vertebrates on Earth, with hundreds of trillions or more of them swimming in the twilight zone.

Undersea lights

The animals of the twilight zone have evolved adaptations to survive in the darkness, including bioluminescent light produced by a chemical reaction (see pp.286–287) that can be used to lure, stun, or confuse prey.

Bristlemouths have evolved rows of light-producing organs called photophores along their undersides, which provide camouflage to help hide from predators and prey. Lantern fish have glowing blue-green lights embedded all over their bodies, which are partly used for signaling and seeking a mate. The firefly squid (*Watasenia scintillans*), meanwhile, produces a bright blue light that it controls with its nervous system, which flashes the animal's bioluminescence on and off.

Crest made up of several elongated branches or "rays" at the start of the long dorsal fin

Giant Oarfish

Regalecus glesne

The giant oarfish is a true ocean giant, and one of the most unusual looking fish. At up to 26 ft (8 m) long, it is thought to be the world's longest bony fish.

Elongated, modified fins resembling oars are thought to be sense organs

The giant oarfish has mirrorlike silver skin speckled with dark spots, and a bright red, body-length dorsal fin. Adults are usually at least 10 ft (3 m) long, and lengths of 26 ft (8 m) have been recorded, with reports of even longer specimens measuring up to 36 ft (11 m).

A member of the bony fish (osteichthyes), the giant oarfish has a flattened and ribbonlike body that makes it resemble a sea snake or eel, but it is not a member of the eel family. It has a long red crest on its head and two modified fins under its head that form the delicate, slender "oars" that give the fish its name. However, giant oarfish do not seem to use their oars like paddles to help them swim. Instead, they may be sense organs that can detect the taste or smell of prey nearby.

Vertical feeder

The giant oarfish is found around the world, except in the colder waters around the poles. It mainly swims in deep water in the open ocean, from about 650 ft (200 m) below the surface down to around 3,300 ft (1,000 m). There, it feeds on plankton, krill, and small fish. Despite its fearsome-looking face, it is not a fierce hunter, and has no teeth, instead filtering its small prey through bony structures in its gills. It is also known for its unusual hunting behavior, hanging vertically in the water and moving up and down. This is believed to be a form of camouflaging while hunting its prey.

Doomsday fish

When a giant oarfish does venture to the surface of the water, it is usually after a storm has passed through, and can mean the oarfish is disoriented, unwell, or even dying. This species' striking appearance and huge size have inspired many nicknames, superstitions, and folk tales in cultures around the world. For herring fishermen, its crest resembled a crown, and they named it the "king of herrings," believing it led shoals of fish through the ocean and could indicate where they were. In several parts of the world, an oarfish washed ashore or spotted at sea is a bad

Crest of the waves

This anatomically accurate 19th-century illustration of an oarfish is taken from *The Animal Kingdom*, by French naturalist and zoologist Georges Cuvier.

Silver star

A diver swims alongside a giant oarfish off the coast of Amami Ōshima island near the southern tip of Japan, helping give a sense of the creature's remarkable proportions. Images like this are unusual, because giant oarfish usually swim at much greater depths.

GIANT OARFISH

> " No fearsome monster of the deeps,
> it turned out to be one of the
> rarest and most beautiful fish
> in the sea—an oarfish. "
>
> BOYD WALKER, zoologist quoted in
> *Oceanography* **magazine, 2014**

omen, foretelling a natural disaster of some kind. This gives it another of its nicknames, the "doomsday fish."

In Japan, the oarfish is known as *ryugu no tsukai*, meaning "messenger from the sea dragon god's palace." Resembling the water dragons of east Asian mythology, oarfish were traditionally seen as the servants of Ryūjin, the dragon god of the sea. They were said to appear at the sea surface or on the shore to warn humans of forthcoming earthquakes. There were several sightings of giant oarfish in the months leading up to Japan's devastating 2011 earthquake and tsunami, leading scientists to wonder if they can sense undersea vibrations before a quake happens—but this has not so far been proven.

Serpent tales
The giant oarfish may have inspired tales of monstrous sea serpents, such as this 1886 illustration based on a sea monster featured on a 16th century map.

White spots on red skin are distinctive markings for this species

Arms contain taste sensors and are lined with suckers that can grasp prey

Red color becomes brighter when octopus is alarmed or annoyed

White-spotted octopus
Most common in the Atlantic, the white-spotted octopus (*Callistoctopus macropus*) is a medium-size octopus, growing up to 5 ft (1.5 m) in length. Its middle two arms are much longer than the other six.

Strolling the seabed
The veined, coconut, or margined octopus (*Amphioctopus marginatus*) hides in other animals' empty shells, and can also "walk" on the seabed using two or four of its arms.

An octopus can squeeze through a hole about the size of its eyeball

Octopuses

Octopoda

Octopuses are highly intelligent animals related to slugs and snails. With eight arms, nine brains, and three hearts, they are able to use tools, build walls, squirt ink, and change color.

There are more than 300 species of octopuses. They are carnivorous cephalopods found in all the world's oceans in a huge range of habitats and depths, from coral reefs to deep-sea hydrothermal vents. Octopuses have a distinctive body structure, with eight arms arranged around a hard, venomous beak; the arms have adhesive suckers on their undersides. Behind this is the head, with two highly developed eyes, and a bag-like abdomen, or mantle. Octopuses can suck in water and squirt it out through a tube on the mantle called the siphon, allowing it to propel itself through the water. Apart from its beak, an octopus's body is soft, and it can squeeze itself through small gaps. Octopuses range in size from the star-sucker pygmy octopus (*Octopus wolfi*), no larger than a couple inches, to the giant Pacific octopus (*Enteroctopus dofleini*), which can reach more than 13 ft (4 m) in length.

Myths of the deep
This 19th-century engraving shows an exaggerated giant Pacific octopus grappling with a French ship. Octopuses and squid have inspired tales of sea monsters for centuries.

BENEATH THE WAVES

Smooth skin can also be arranged into branching, treelike spikes for camouflage

Octopuses have the ability to disguise themselves by altering their color and shape. A common octopus (*Octopus vulgaris*) can transform from its usual smooth, white coloration and shape into a knobbly, speckled, green and brown clump in less than a second, enabling it to camouflage itself among seaweed and corals. This camouflage helps octopuses hide from predators such as sharks and dolphins, and to hunt their own prey, which includes sea snails, clams, fish, and crabs. The mimic octopus (*Thaumoctopus mimicus*), first discovered off an Indonesian island in 1998, can even imitate other animals, arranging its body to resemble a venomous lionfish (*Pterois*) or sea snake (see pp.264–267) to deter predators. In common with some other cephalopods, octopuses can also squirt an ink, made of mucus and melanin pigment, as a defensive mechanism when threatened.

As well as a central brain, octopuses have a sub-brain for each arm, making nine brains in total, and studies have shown them to be highly intelligent animals. Individuals have learned how to unscrew jars to reach food inside them, solve mazes, and escape from closed containers. In captivity, they have been observed playing with toys—a sign of intelligence in animals—pushing them around or arranging them in a pattern. They can recognize different humans, sometimes squirting a jet of water at people they do not like. They often find ways to escape from aquarium tanks, even climbing from one tank into another to look for prey.

Octopuses can also use tools, another behavior indicating high intelligence usually only found in vertebrates such as mammals and birds. Some species arrange seashells or pebbles around their dens, or even use them to build small walls.

Patchwork patterns
The pale octopus (*Octopus pallidus*) is covered in a mosaic of small patches and, like many octopuses, can change both its color and its texture.

The veined octopus (*Amphioctopus marginatus*) has been observed sheltering in empty shells or other containers, and has been seen carrying coconut shells as it "walks" on its tentacles across the seabed. If danger threatens, it will stop and quickly hide in the shell.

The intelligence of the octopus has been widely studied and is somewhat anomalous—problem-solving and planning are unusual in invertebrates. Their complex nervous system is not yet fully understood. Octopuses are also unusual among intelligent animals because they are short-lived, surviving just a few years on average—most animals with comparable intelligence live longer.

Water vessel
This octopus vase was made in Knossos, Crete, and dates from c. 1500 BCE. Octopuses often feature on artifacts from Crete's Minoan culture.

Suckers on tentacles are depicted accurately

> " I took several specimens of an Octopus, which possessed a most marvelous power of changing its colors; equaling any chameleon. "
>
> CHARLES DARWIN, letter to John Stevens Henslow, 1832

Gods and stories

Octopuses are part of the folklore and culture of several coastal and island nations. The nuu or octopus often features in the art and stories of the Haida people of the Pacific Northwest, who admire it for its cunning and flexibility. In Fijian folklore, a fierce octopus goddess, Rokobakaniceva, defeated the shark god Dakuwaqa in battle. Meanwhile, in the legends of the Ainu people of northern Japan, Akkorokamui is an enormous octopus-god, said to lurk in Uchiura Bay on the island of Hokkaido.

QUICK-CHANGE COLOR

Octopuses change color using specks in their skin called chromatophores. Each chromatophore contains a flexible sac of pigment, and is surrounded by a ring of muscles. When the muscles contract, they pull the sac outward, creating a large spot of color. When they relax, the sac shrinks to a tiny speck. The octopus's brain can control thousands of chromatophores in coordinated sequences to create changing colors and patterns.

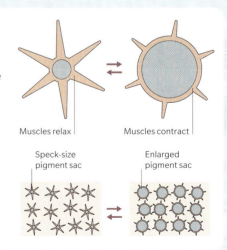

Muscles relax | Muscles contract
Speck-size pigment sac | Enlarged pigment sac

CHROMATOPHORES

BROWN PAPER NAUTILUS
Argonauta hians
Also known as winged argonauts, females of these small octopuses make a spiral-shaped, papery case around themselves to hold their eggs.

GREATER BLUE-RINGED OCTOPUS
Hapalochlaena lunulata
This tiny octopus is one of the deadliest creatures in the sea due to its venomous bite. When alarmed, it flashes a warning in the form of blue rings on a yellow background.

DUMBO OCTOPUS
Grimpoteuthis
Grimpoteuthis is a group of deep-sea octopuses, found up to 23,000 ft (7,000 m) below the surface. They are known as "dumbo" octopuses because of their large, elephant-like "ears."

Sail shaped like a crab claw

Pacific voyager
This model recreates a Lakatoi, a traditional multihulled sailing boat from Papua New Guinea.

Pacific Ocean

The Pacific is the largest expanse of water on Earth. It covers almost one-third of the planet's surface, and takes up more space than all the world's land combined. It is also home to the deepest point in the ocean, and more than 20,000 islands.

Lying with Antarctica to the south and the Arctic to the north, and between the Americas to its east and Asia and Australasia to its west, the Pacific encompasses 60 million square miles (155.4 million sq km).

It contains at least 200,000 named marine species, including blue whales and humpback whales, the Pacific white-sided dolphin, great white and hammerhead sharks, sea turtles, and many species of penguin. Scientists believe there are hundreds of thousands more species awaiting discovery, many of them in the ocean's little-explored depths.

The first explorers

Early humans began to spread out eastward from Asia to settle on Pacific islands as long as 30,000 years ago, crossing the water on dugout

Maris Pacifici map, 1589
This map by Flemish mapmaker Abraham Ortelius, though not fully accurate, was the first map of the Pacific to be printed. It includes an illustration of Magellan's ship *Victoria*, being guided across the ocean by an angel.

canoes or simple rafts. By 3,000 years ago, outriggers or double-hulled canoes, with two connected hulls to increase stability, allowed explorers to travel even greater distances across the open ocean. Over the millennia, a variety of seagoing and fishing cultures evolved on the thousands of islands scattered across the Pacific, collectively known as Oceania. It is not known who first sailed all the way across the Pacific Ocean in one journey. Some theories suggest it could have been ancient Polynesians on trading voyages, or Japanese fishing boats sailing off course, possibly ending up as far away as the coast of South America. In recorded history, this famous first was claimed by Portuguese explorer Ferdinand Magellan, who set off from Spain in 1519 to find a new westward trading route to the spice islands of Indonesia. After passing through

260 BENEATH THE WAVES

Pearl divers
This 1836 woodblock print by the Japanese artist Hokusai, shows the Ama female divers at work while men wait in boats to retrieve their catch of pearls or abalone shellfish.

Feathers adorn the figure's neck

Karamanua figure
Carvings of the sea spirit Karamanua, such as this one from the Solomon Islands, were built into boathouses and given offerings in the hope of good luck when fishing.

the stormy straits at the southern tip of South America, Magellan hugely underestimated the scale of the Pacific, expecting to reach Asia in just a few days. Instead, it took more than three months. Running out of supplies, 19 crew members died of scurvy on the crossing.

Peaceful ocean

Magellan named the Pacific Ocean. Compared to the rough waters of the straits he had sailed through, the ocean was calm, so he called it the *Mar Pacifico*, or "Peaceful Sea." However, it is no calmer than any other ocean basin. Enormous breakers are common—the biggest rogue wave on record was measured at 58 ft (17 m) off Vancouver in 2020—and the western Pacific in particular has powerful tropical cyclones. The Pacific's climate is tropical close to the equator: hot, humid, and rainy. Elsewhere, it is subtropical, with hot summers and mild winters. Earthquakes and volcanoes, common around the ocean's edges, can set off tsunamis.

Many Pacific peoples are expert fishers, boat-builders, and navigators. Crafts and workmanship are also a feature of Pacific cultures. On the island of Malaita in the Solomon Islands, some people still use a traditional form of money called *tafulie*, made from seashells carved into small disks and threaded onto long strings. Many cultures used shells and sharks' teeth to make fish hooks, jewelry, and ceremonial costumes. Razor-sharp sharks' teeth were used to make the serrated edge of the *leiomano*, a traditional Hawaiian fighting club, as well as being used as cutting and carving tools.

Along the Pacific's western coast, female divers such as the Ama of Japan and the Haenyeo of Korea's Jeju Island collect shellfish, pearls, and seaweed. Holding their breath as they plunge to depths as great as 33 ft (10 m), they can remain submerged for up to two minutes. The nomadic Bajau of Malaysia and the Philippines have been deep-sea spearfishers for so long they have evolved larger spleens to store extra oxygen. Some can hold their breath for more than five minutes and dive to depths of 200 ft (70 m).

Ocean folklore

Pacific myths, legends, and folktales are full of sea serpents, monsters, dragons, mermaids, and water spirits. In Fijian folklore, half-shark, half-man sea god Dakuwaqa protects the islands and their fishing boats. In Chinese legend, the Jiaoren are mermaids who weave fabric from sea silk. Their tears turn into pearls, which can grant humans the ability to breathe underwater.

Many Pacific Ocean gods and spirits can control the weather and fish. Karamanua of the Solomon Islands is a shark-headed sea spirit who rides rainbows, overturns boats, and shoots flying fish as arrows—but also helps humans

Catching waves
This illustration of Hawaiian surfers dates from 1855. Surfing was a traditional pastime in Hawai'i and Polynesia, but missionaries in Hawai'i discouraged it and it dropped in popularity until a revival in the early 20th century.

" The call of the waves is always stronger than the call of duty. "

DUKE KAHANAMOKU, *famous Hawaiian surfer and swimmer, 1890–1968*

by sending shoals of fish and fine weather. In Hawaiian mythology, shape-shifting, lizard-like water guardians called Mo'o take the form of beautiful women. One, Kalamainu'u, is said to help fishers catch wrasse. She assumed the shape of a human to go surfing, where she caught the eye of a young chief, Puna, and lured him to her cave.

Dating from around 400 CE, surfing features in old Hawaiian stories. Polynesians brought a form of bodyboard surfing to Hawai'i, where it developed into the art of standing upright while riding huge crested waves.

GREAT PACIFIC GARBAGE PATCH

Like the Atlantic, the Pacific has two huge circulating surface currents, or gyres—one north and one south of the equator. In the North Pacific gyre, disintegrating plastic waste, thrown into the sea or into rivers around the Pacific, has collected into two swirling masses. Together, they are known as the Great Pacific Garbage Patch. Scientists are working on clean-up devices and methods to find an effective way to rid the Pacific of its plastic problem.

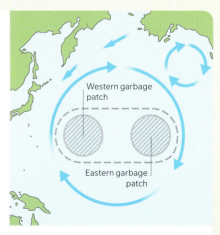

GARBAGE PATCH FORMATION

Hammerhead Sharks

Sphyrnidae

Known for their T-shaped heads, hammerhead sharks are among the most recognizable fish in the sea. Their distinctive "hammer" is an adaptation that helps them track down prey more easily.

School of hammerheads
Hammerheads such as these pictured off the Galápagos Islands can gather in huge schools of hundreds of individuals.

Cephalofoil has smoother, more rounded shape than that of other hammerheads

Tail fin is asymmetrical

Smooth hammerhead
Unusually among large hammerhead species, smooth hammerheads (Sphyrna zygaena) do not have notches along their heads and prefer temperate waters.

Clearly identified by their hammer-shaped heads known as cephalofoils—a compound of the Greek words for head and wing—there are around ten species of hammerhead. The largest is the great hammerhead (Sphyrna mokarran), which can grow to around 20 ft (6 m) long. They live in temperate and tropical oceans, but can tolerate colder waters when they dive down to hunt prey. Its head shape allows the shark to move and turn quickly in the water, and it can move its head up and down in a way that many other sharks cannot.

Sensing, feeding, and breathing

With eyes set on either end of the cephalofoil, a hammerhead has a wide field of vision and good depth perception. Turning its head from side to side as it swims allows the shark to have 360-degree vision.
Sharks have electrical sensors called "ampullae of Lorenzini" on their heads. The cephalofoil provides a larger surface area for these sensors, so it makes hammerheads more sensitive to the electrical signals given off by prey, enabling them to find their common prey—stingrays—hidden in the sand more easily.

Hammerhead sharks also feed on octopuses, squids, fish, and crustaceans. They have small mouths with sharp teeth at the front and flat, grinding teeth at the back to help them crush through animals with hard shells, such as crabs and lobsters.

Like most fish, hammerheads breathe through their gills. In cold water, this can cause them to lose heat quickly. Scalloped hammerheads (Sphyrna lewini) have been recorded stopping gill irrigation while diving to the deep sea—where waters are much colder—which enables their bodies to remain warm enough to hunt.

In Māori culture, sharks are respected and revered. The mangōpare is a curved symbol used in traditional Māori art that represents the hammerhead and stands for strength and determination.

Jaws below
This 1894 illustration of a smooth hammerhead shows the position of its five gill slits near its pectoral fin.

BONNETHEAD
Sphyrna tiburo
One of the smaller hammerhead species, bonnetheads are known for their shovel-shaped cephalofoils.

SCALLOPED HAMMERHEAD
Sphyrna lewini
Scalloped hammerheads can be identified by the notches than run along the edge of their hammer.

" *Kaua e mate wheke. Me mate ururoa.* "
(" Don't die like an octopus.
Die like a hammerhead shark. ")

MĀORI PROVERB

BENEATH THE WAVES

Modern **sea snakes** evolved around **10-15 million** years ago, when some venomous **snakes** began living in the sea

Striped and deadly
The black-banded sea krait, or Chinese sea snake (*Laticauda semifasciata*), is one of the most venomous sea snakes. It rarely attacks humans, mainly using its venom to immobilize prey.

Vivid stripes are found on many species of sea snakes

BEAKED SEA SNAKE
Hydrophis schistosus
The aggressive beaked sea snake grows to about 4 ft (1.2 m) long. It is often found in shallow waters or in estuaries, and also sometimes on muddy beaches.

YELLOW-BELLIED SEA SNAKE
Hydrophis platurus
This smallish snake, about 30 in (80 cm) in length, is instantly identifiable by its black and bright yellow markings. It is wide-ranging, often found in the open ocean.

Sea Snakes

Sea snakes are saltwater reptiles that have adapted to life in warm seas; they were originally part of the cobra family. Most give birth to live young in the water and, like their land-based relatives, most of the species are venomous.

Because sea snakes are reptiles, they have to come to the surface to breathe air. Some species have been spotted making dives as deep as 820 ft (250 m) in search of prey, usually small fish. Most of the time they are found in shallower waters, often around coral reefs and on sandy seabeds, in the Indian and Pacific ocean regions.

Sea snakes are fully adapted to life in the water. They have narrow bodies and flattened, paddlelike tails, which they move from side to side to power themselves through the ocean. A sea snake's single lung extends almost the whole length of its body, allowing it to take in and store a lot of oxygen. This means it is buoyant when on the ocean surface, and when it swims underwater, the weight of the water compresses its lungs, so it reaches neutral buoyancy to allow it to swim with ease. Although they do not have gills, sea snakes can absorb extra oxygen from the seawater through their skin. A few species sometimes crawl on to land, but they can only move slowly and awkwardly there, because their bodies have evolved for water.

Powerful venom
Some of the most venomous of all snake species are sea snakes, including the black-banded sea krait (*Laticauda semifasciata*) and the extremely deadly Dubois' sea snake (*Aipysurus duboisii*). However, sea snake bites and deaths are rare because most species are timid and reluctant to strike, even when handled. Fishermen often catch sea snakes in their nets by accident, and simply untangle them and throw them back without being bitten. One species is known for its aggressive temperament: the beaked sea snake (*Hydrophis schistosus*). If caught in a fishing net, it will fight back, and is responsible for most sea snake bites and fatalities.

SEA SNAKES

THE GREAT SEA SERPENT
(according to Hans Egede.)

For centuries, sailors and seafarers have reported sightings of huge sea serpents, big enough to attack ships and swallow people whole, but as far as is known, there are no giant sea snakes in the oceans. Compared to land snakes, sea snakes are small, with even the biggest species, the yellow sea snake (*Hydrophis spiralis*), rarely reaching 10 ft (3 m) in length. However, 50 million years ago, an enormous "sea serpent" did exist: *Palaeophis colossaeus* may have reached 40 ft (12 m) in length. It belonged to an earlier group of aquatic snakes, which died out before today's sea snakes evolved.

Fake snakes

Fossils of such ancient monsters could have inspired some accounts of real-life sea serpents with monstrous proportions, but there could also be other explanations. While sea snakes do not grow very big, some other long, skinny, snake-shaped sea creatures do. The giant oarfish (*Regalecus glesne*) is an eel-shaped ocean fish that can reach 36 ft (11 m) in length (see pp.248–249). It has a red, dragonlike crest on its head, similar to those often seen in sea serpent depictions. Giant pyrosomes and siphonophores are another

Greenland sighting
This engraving from 1839 is based on a report of a giant sea serpent made by missionary Hans Egede. He described seeing a "most dreadful monster," bigger than a ship and with snakelike features, emerge from the sea near Greenland.

" On the Coasts of Norway ... do all agree in this strange story, that there is a Serpent there which is of a vast magnitude, namely 200 foot long, and more – over 20 feet thick. "

OLAUS MAGNUS *Carta Marina*, 1539

possibility. These are both colony animals, made up of hundreds of smaller creatures living as a community. They can glow with their own light, and can reach lengths of 60 ft (18 m) and 130 ft (40 m) respectively. Myths and legends, religions, and creation stories from all over the world feature sea snake gods and goddesses. They sometimes represent the sea itself, or the forces of chaos and evil. Some are giant snake-fish or water snakes, while others, especially those from Asian legends, are better described as snake-shaped water dragons.

In West African and Haitian creation myths, Damballa is a giant serpent-god who created the whole sea by shedding his skin, and can appear to humans as a rainbow. In ancient Greek mythology, the Cetus was a type of snakelike or eellike sea monster. It was slain by the hero Perseus when he came to rescue Andromeda after she was tied to a rock and left as a sacrifice by her father, King Cephus. In Japanese mythology, the ocean-protecting god Ryūjin is an enormous serpent-dragon who lives in a red-and-white coral palace on the seabed. He controls the tides, rain, and thunder, and is the king of all snakes.

The largest legendary sea serpent must be Jörmungandr, the Miðgarðsormr, or "World Serpent" of Norse mythology. The eternal enemy of the thunder god Thor, this monstrous snake lies in the sea encircling the whole world, holding his tail in his mouth.

Syrian serpent
A dangerous sea serpent said to live in the Mediterranean Sea near Syria depicted in an edition of *Marvels of Creatures and Strange Things Existing*, originally written by Zakariya al-Qazwini in 1280.

Yoshisada and Ryūjin
This 19th-century carved plate depicts Nitta Yoshisada, a real samurai and clan leader from 14th-century Japan, praying to the sea serpent god Ryūjin before going into battle.

In parts of Japan, sea snakes are hunted for food. Despite their deadly venom, experienced hunters catch them by hand

By sea and land
Unusually for a sea snake, the yellow-lipped sea krait (*Laticauda colubrina*) spends around half of its time on land. It hunts in the sea, but comes ashore to rest, mate, and lay its eggs.

Speed and grace
Reaching speeds of up to 68 mph (109 kph), Atlantic sailfish (*Istiophorus albicans*) are among the fastest known fish in the ocean.

Huge, sail-like dorsal fin gives sailfish their name

Bill is used to attack prey such as sardines

Billfish

Istiophoridae, Xiphiidae

Known for their distinct swordlike bills, these fish are effective predators. Rather than spearing their prey, they slash their bills through the water to stun or injure smaller fish.

Marlin, sailfish, and spearfish (Istiophoridae) and swordfish (Xiphiidae) are all types of billfish—large, predatory saltwater fish with a long spear-like rostrum, bill, or beak, which they use to hunt prey. These enormous animals are often apex predators in their habitats. They can grow to several yards long, and weigh more than 1,000 pounds; the heaviest swordfish caught in competition weighed 1,182 lb (536.15 kg). Swordfish also have the longest bills, which can be up to one third of their body length.

Camouflage and hunting
Billfish are generally blueish silver with a lighter belly and darker back—different species have slightly different coloration and patterns. This countershading helps provide them with camouflage in the water above and below them. They are found all around the world in temperate, subtropical, and tropical waters, and are highly migratory, swimming thousands of miles across the seas. They also migrate vertically up and down the water column and can dive to around 10,000 ft (3,000 m)—the edge of the abyssal zone (see pp.298–299)—to hunt deep-sea animals, such as squid, and also to hide from their own predators. This is enabled by a sophisticated swim bladder that allows them to adjust to pressure changes They hunt by using their sword-like bills to slash at prey, and wound or stun them before eating.

BILLFISH 269

Skin color changes with excitement. Scientists have observed this species darkening to almost black before attacking a bait ball

SHORTBILL SPEARFISH
Tetrapturus angustirostris
This marlin, found principally in the Indian and Pacific Oceans, has a much shorter rostrum than other billfish. Its bill is only slightly longer than its lower jaw.

STRIPED MARLIN
Kajikia audax
Striped marlin hunt in groups in tropical and subtropical waters, and can increase the contrast of their stripes just before an attack.

Billfish have long featured in myth and literature. In Singaporean folklore, the area of Bukit Merah, known as Redhill, was once under attack from swordfish, so the king's men went into the water to fight them. Realizing they were no match for the fish, a boy suggested they put banana tree trunks into the water so that the swordfish would injure themselves on the wood. This worked, but in a jealous rage the king ordered the boy to be killed. The boy's blood flowed until the hill was stained red. Herman Melville wrote about swordfish stabbing ships in his novel, *Moby-Dick*. Such stories have a basis in real billfish behavior—in 1967, an 8-ft- (2.4-m-) long swordfish impaled itself in the side of the research submersible ALVIN. The crew had to resurface and remove it.

Two swordfish
This 1923 painting was made by scientific illustrator Hashime Murayama for the National Geographic Society.

Bivalves

Bivalvia

Known for producing pearls, bivalves take their name from the shape of their shells, which have two parts (valves) joined by a hinge.

The **Asian moon scallop** can swim at up to 18 in (45 cm) per second

First appearing more than 500 million years ago, bivalves now include an estimated 106 families comprising more than 9,000 species, including oysters, scallops, clams, cockles, and mussels. Found around the world, from river beds and shorelines to the deepest ocean, and from Arctic to tropical waters, most bivalves live at the shoreline or in shallow water. However, a few, such as the giant mussel (*Bathymodiolus thermophilus*), live around hydrothermal vents in the Pacific Ocean at depths of around 9,200 ft (2,800 m). Here, they derive nutrients from the bacteria that live in their gills and subsist on chemicals from the vents (see pp.294–295).

Open and shut

Bivalves' distinctive hinged shells are made from calcium carbonate, and formed by secretions from an underlying layer of tissue called the mantle. The hard outer

Bivalves in art
Venus rises from the waves while standing on a scallop shell, an ancient Roman symbol of female sexuality, in Sandro Botticelli's painting *The Birth of Venus* (c. 1485).

Valves open during the day to expose the algae to sunlight

Colossal clams
Giant clams (*Tridacna gigas*) grow up to 4 ft (1.3 m) long due to algae called zooxanthellae that live inside their shells. The algae get protection, while the clams get nutrients from the algae.

Bright colors in the body tissues result from the algae living within them

surface may be roughened like an oyster's, ribbed like a scallop's, or feature concentric ridges like a clam's. The inner surface, by contrast, is smooth; this layer is formed from an iridescent substance called nacre or mother-of-pearl that grows in some mollusks.

The two valves are held together by an elastic ligament, which allows the shell to spring open. Powerful muscles in the bivalve's body act to close it. Scallops use this opening-and-closing action to swim—opening their valves to create a vacuum that draws water inside the shell, then using their muscles to "clap" the shell shut, pushing the water out through vents near the hinge and shooting it forward.

> " What pearl art thou that none possesseth the price of thee? "
>
> RUMI (JALAL AD-DIN MUHAMMAD AR-RUMI),
> *Divani Shamsi Tabrizi,* c. 1247 (English translation 1898)

Most bivalves, however, live their adult lives in one place. Some, such as mussels and oysters, secrete tough proteins or a cement-like substance to anchor themselves to rocks. One species, the noble pen shell (*Pinna nobilis*), secretes long, fine threads that were historically valued as "sea silk" and woven into cloth. Others use their muscular foot to burrow into river and sea beds.

SCALLOP
Pectinidae

Found throughout the world's ocean, scallops have simple eyes at the edge of their mantles.

HOOKED MUSSEL
Ischadium recurvum

Inhabiting the North American Atlantic coast, these bivalves are identified by their claw-shaped shells.

THORNY OYSTER
Spondylus

This bivalve was used by the Inca and other Andean peoples in rituals and was made into jewelery.

Most bivalves are filter feeders, drawing water into the shell and using their gills to filter out food particles such as phytoplankton. They are also an important food source for many other species, from fish and sharks to seagulls, sea otters, walruses, and humans. A few species are carnivorous, such as the clam *Poromya granulata*, found in the north Atlantic, which catches small crustaceans. The giant clam has a symbiotic relationship with the algae that live in its body tissues—they supply the clam with 65 to 70 percent of its nutrients.

Some species clean the water in which they live, taking in toxins from algae, sewage, or particles of heavy metals. This tendency to absorb toxins can make them poisonous to other animals that eat them—including humans.

Tools and pearls

For thousands of years, people in coastal areas around the world have eaten and used bivalves. Today, commercial farming of bivalves is big business: the National Oceanic and Atmospheric Administration (NOAA) estimated that in 2011 153.6 million lb (69.67 million kg) of bivalves were harvested in the US alone.

The indigenous peoples of the Palau islands and Guam in the Pacific Ocean used clam shells to make tools, and as repositories for valuables or ancestors' skulls. Giant clams feature in the creation myth of the Palauan people, while in Tahitian legend the giant clam Pua Tu Tahi is a fierce god of the sea. In Europe, the scallop shell was a symbol of the Greek goddess of love, Aphrodite (Roman equivalent was Venus), and signified fertility. The shell was later adopted by Christians as an emblem of the apostle James and worn by medieval pilgrims traveling to his shrine at Santiago de Compostela in Spain, as well as featuring on medieval churches and other buildings.

The most prized products of bivalves, however, are pearls. If an irritant such as a piece of debris enters a bivalve's shell, it will increase its secretion of nacre to surround it. Over the years, the lump grows to become a pearl. Most bivalves can create pearls, but oysters and freshwater mussels produce the most lustrous ones. In the Persian Gulf, pearl-diving was a lucrative business for thousands of years before oil was discovered. Pearls were harvested by divers descending up to 65 ft (20 m), using only wooden nose clips and weights tied to their legs. Known as "tears of the sea," pearls were highly prized among the peoples living in the Gulf and were exported as far as China and western Europe from around the 6th century CE onward.

Shell opens along the longest side, like a book

Razor's edge
These razor clams (*Ensis leei*) hide from predators by quickly burrowing into wet sand, leaving just a breathing hole at the surface. They can be caught by squirting salty water into the hole, which makes the clam emerge.

Clam nectar, or clam juice, is made by steaming fresh clams

Key ingredient
Concentrated clam broth, such as this bouillon from the 1920s, is used in seafood recipes such as clam chowder.

Bivalves and other animals
This page from *The Genera Vermium* (1783) by English naturalist and artist James Barbut shows different species of bivalves, as well as chitons (top left, 1) and barnacles (top center left, 3).

BENEATH THE WAVES

Basque fisheries
The Basque Country is known for its tuna fishing, which is promoted in this 1940 tourism poster by Julio García.

BIGEYE
Thunnus obesus

Bigeyes look similar to yellowfins (see below), but they have—as the name suggests—larger eyes. They also have black edges on their finlets.

SOUTHERN BLUEFIN
Thunnus maccoyii

Although population numbers are starting to recover, this species is categorized as endangered by the IUCN—a result of overfishing since the 1950s.

YELLOWFIN
Thunnus albacares

This striking species can be identified by its bright yellow dorsal, anal, and tail fins, and yellow finlets.

PAYS BASQUE

Tuna

Thunnini

These powerful predators are among the ocean's fastest fish, but are also preyed on by other ocean animals—and humans. Overfishing has almost driven some populations to extinction.

Tuna are painted in the center of the bowl

Found in warm seas worldwide, tuna grow to around 6½ ft (2 m), but the largest species—the Atlantic bluefin tuna (*Thunnus thynnus*)—can reach 13 ft (4 m) long. Tuna eat other fish, squid, and crustaceans, and are preyed upon by large predators such as sharks and toothed whales.

These ocean giants are built for speed. Their hydrodynamic, streamlined bodies allow them to reach around 45 mph (28 kph), making them one of the fastest fish in the sea, alongside mako sharks (*Isurus*), certain billfish (see pp.268–269), and others. Their scientific name—*Thunnus*—probably comes from the ancient Greek word *thynō*, meaning "to rush."

Despite their size, speed, and hunting prowess, many tuna species are under threat. In the UK, Atlantic bluefin tuna stocks almost collapsed due to overfishing. After a recovery plan was put in place in 2007, populations started increasing and the species was removed from the International Union for Conservation of Nature (IUCN) endangered species list in 2021.

The king of sushi

We know from the writings of ancient Greek, Roman, and Arab scholars that tuna has been fished since antiquity. Today, it is widely used in sushi, but this was not always the case. In the Japanese Edo period, tuna was unpopular: these huge, powerful fish were difficult to catch and the meat spoiled quickly. It was also considered unlucky because its name, *shibi*, sounded like the word for the "day of death." Over time, better preservation and refrigeration methods made it easier to keep longer and, around 1830, a huge catch lowered prices significantly so more people started eating it.

Tuna meat is sometimes known as the "king of sushi." Tokyo's Tsukiji fish market is famous for selling tuna for huge prices at its annual New Year's auctions. In 2019, a 613 lb (278 kg) bluefin tuna sold for $3.1 million at a Japanese auction, making it the world's most expensive tuna.

Tuna are found in every ocean and are important in many cultures. The Polynesian goddess Kohara, the goddess of tuna, is said to be the ancestor of all tuna fish. In the Philippines, the city of General Santos holds an annual festival with a parade and cultural performances to celebrate its thriving tuna industry.

Fish bowl
This ceramic bowl—probably of Spartan origin—has a rosette pattern of tuna encircled by dolphins. Its creator is known only as "The Fish Painter."

Crossing oceans
Tuna often migrate in large shoals. They are known for their incredible ability to travel thousands of miles across the ocean.

A shiver of sharks
Above one of the many shipwrecks that litter the coast of North Carolina, a group of sand tiger sharks (*Carcharias taurus*) congregates below a school of Atlantic spade fish (*Chaetodipterus faber*). Sharks are often thought of as fierce and solitary, but sand tigers do not conform to the stereotype. Despite their fearsome appearance, they rarely attack humans and often come together in groups, displaying complex social behavior normally seen in mammals.

Squid

Myopsida, Oegopsida, Bathyteuthida

From the tiny southern pygmy squid (*Idiosepius notoides*), measuring less than 1 in (2 cm), to the colossal squid (*Mesonychoteuthis hamiltoni*) at up to 46 ft (14 m) long, squid are found throughout the ocean.

> The **first photo** of a living **giant squid** in the wild was taken in **2004**

Like the octopus and cuttlefish, squid are a type of cephalopod, a Greek word that translates as "head foot." This name refers to the way the mantle (the pocket of skin covering the body) is directly attached to the arms and tentacles. There are more than 300 different species, including large squid, such as the Humboldt squid (*Dosidicus gigas*), the giant squid, and the colossal squid (*Mesonychoteuthis hamiltoni*) that live in the deep sea, hundreds of yards below the surface. Many smaller species are also found in the shallows, such as the Caribbean reef squid (*Sepioteuthis sepioidea*) and bigfin reef squid (*Sepioteuthis lessoniana*) that live on coastal reefs.

Squid schools
Although some squid are solitary, others, such as these Caribbean reef squid, live in groups of up to 30 individuals.

SQUID 279

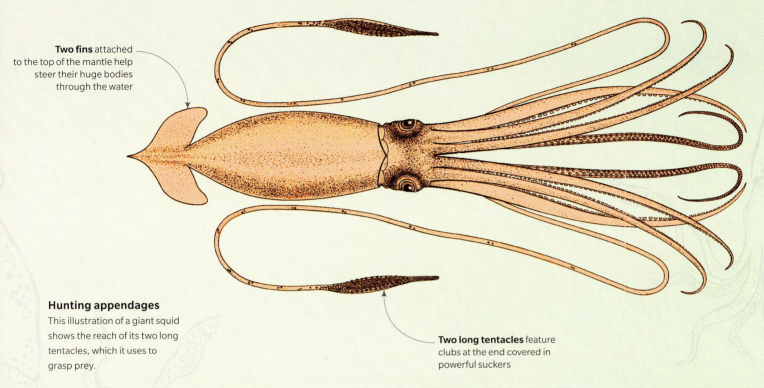

Two fins attached to the top of the mantle help steer their huge bodies through the water

Hunting appendages
This illustration of a giant squid shows the reach of its two long tentacles, which it uses to grasp prey.

Two long tentacles feature clubs at the end covered in powerful suckers

Squid have a sharp beak and a tonguelike organ called a radula, and they have ten flexible limbs—eight arms and two longer tentacles—that are used for swimming and feeding. Their skin contains pigment-filled organs called chromatophores that allow them to rapidly change color and pattern. These can be used for both camouflage and communication.

Squid have an internal shell called the gladius, or pen, which protects its organs and provides a place for muscles to attach to. Because they have no external shell, squid can be vulnerable to predators and some species have developed sophisticated self-defense strategies. When the octopus squid (*Octopoteuthis deletron*) feels under threat, it distracts its predator long enough to get away by detaching a bioluminescent arm and leaving it behind.

Elusive giants
One species of squid dominates myths and legends: the giant squid, which lives in waters hundreds of yards deep and is rarely seen by humans. Scientists believe the giant squid's eyes—among the largest in the animal kingdom—have developed to this size to give it a better chance of seeing predators approaching in the dark ocean. However, this does not mean it can avoid conflict. It is known to fight with sperm whales (*Physeter macrocephalus*), and some of these whales have been seen with sucker scars on their bodies and squid beaks have been found in their stomachs.

Squid attack
Even though giant squid are not believed to target ships or humans, the species has inspired stories of sea monsters for hundreds of years. The Caribbean Lusca and the Scandinavian Kraken, may both have been based on giant squid sightings. Jules Verne's classic 1869 novel *Twenty Thousand Leagues Under the Sea* is known for the ferocious battle between the crew of the exploration vessel the *Nautilus* and a giant squid.

Squid is depicted in mother-of-pearl

Curious case
This early 19th-century Japanese *inrō* (case) created by lacquer artist Hara Yōyūsai features a design depicting a squid, shell, and seaweed.

EUROPEAN SQUID
Loligo vulgaris
Growing up to 16 in (40 cm) long, the European squid can be found as deep as 1,600 ft (500 m).

BIGFIN REEF SQUID
Sepioteuthis lessoniana
This is a rare example of a squid that displays paternal care. Males investigate nesting sites before the female lays her eggs.

FIREFLY SQUID
Watasenia scintillans
Also known as the sparkling enope squid, this tiny squid is named for its glittering blue bioluminescence.

Coral cohabitants
A shoal of ribboned sweetlips (*Plectorhinchus polytaenia*) are joined by a lone humpback red snapper (*Lutjanus gibbus*) in the warm, shallow waters around a reef in the Indian Ocean.

The ribboned sweetlips fish is native to the Indian Ocean and the western Pacific

Elongated dorsal fin that resembles bat wing gives the fish its name

Silver, gray, or brown skin on body

Indian Ocean

The third-largest of the world's five great ocean basins, the Indian Ocean is known for its warm waters and has a rich history of travel and trade.

The Indian Ocean basin is bordered by Africa, Arabia, and southern Asia to the north, and by the Southern Ocean to the south, making it a predominantly tropical ocean. Unlike the Atlantic and Pacific ocean basins, it does not stretch between both polar regions, and it also has fewer islands and marginal seas than the Atlantic and Pacific.

As a consequence of its mainly tropical and subtropical location, and being cut off from polar waters in the north, the Indian Ocean basin has the warmest average water temperature of all the world's ocean. This means it regularly generates powerful tropical cyclones (see pp.146–147), as well as the planet's strongest wind-and-rain monsoon cycle. The monsoon is caused by the warming up of the Indian subcontinent each summer, creating areas of low pressure that draw in humid air from the Arabian sea. It brings heavy downpours to India's mainland and the surrounding countries, but it also causes rainstorms over the ocean itself at different times of year.

Flowing fins
This 19th-century illustration shows a longfin batfish (*Platax teira*), common in the Indian Ocean near reefs and wrecks.

Bound for Oman
A ship sails the Indian Ocean to Oman in a 13th-century manuscript of the *Maqamat al-Hariri*, a book of tales by al-Hariri of Basra.

INDIAN OCEAN

BENEATH THE WAVES

Large scale migration
Christmas Island red crabs crowd onto the beach to find a mate during their annual migration. The females later lay their eggs in the water.

The Indian Ocean is known for its many coral reefs, especially in the Red Sea and southeast Asia, and for coral atoll island chains such as the Maldives and Comoros. These are home to notable species such as the vividly-colored peacock mantis shrimp (*Odontodactylus scyllarus*), the endangered hawksbill turtle (*Eretmochelys imbricata*), and the huge whale shark (*Rhincodon typus*). Indian Ocean islands are also famous for their wildlife, including one of the world's largest tortoises, the Aldabra giant tortoise (*Aldabrachelys gigantea*) on Aldabra in the Seychelles, and the now-extinct dodo (*Raphus cucullatus*) of Mauritius. In the eastern Indian Ocean, Christmas Island red crabs (*Gecarcoidea natalis*) annually migrate in their millions from inland forests to the coast to mate and breed. The Indian Ocean was also where the first living coelacanth fish (order Coelacanthiformes)—an unusual fish believed to have died out with the dinosaurs 66 million years ago—was discovered in 1938.

Travel, exchange, and people

The Indian Ocean's geography has historically made it easy to travel across, with seafarers able to traverse the whole region while remaining in ice-free waters close to the coast. Here the sea has linked many different peoples around its shores through journeys of exploration and trade, enabling the transfer of languages, people, place names, crop cultivars, and goods such as pottery across the region. Traditional Arabic *dhow* sailing

> " The Indian Ocean is chaos. It is plenitude.
> It is also what we cannot ever fathom. "
>
> MEENA ALEXANDER, quoted in an interview, 2010

Portuguese coat of arms and emblem of Manuel I of Portugal

Navigation aid
Found in a Portuguese wreck off the coast of Oman, this is one of the earliest known marine astrolabes, dating from the late 15th century.

DISPLACED ELEPHANTS

In the 19th and 20th centuries, elephants were shipped to India's Andaman Islands to move heavy tree trunks in logging operations. They were moved between islands by making them swim across the sea. Logging was banned in 2001, but some elephants remained on the islands, both as domesticated animals and in a small feral population.

ANDAMAN ISLANDS ELEPHANT

boats have been journeying up and down the ocean's coasts for more than 1,000 years; the boat's design and name spread to Africa, where it is still used today for cargo, fishing, and tourism. Arabic explorers and traders also sailed eastward, reaching as far as China; while in the other direction, admiral and diplomat Zheng He explored this vast region on several missions from China in the 1400s.

On the large African island of Madagascar, a trading hub over many centuries, groups such as the coastal Cotier peoples draw on a combination of southeast Asian, African, and Arabic heritage, largely brought about by ocean travel. Other groups—such as the Sentinelese of India's Andaman Islands—are among the world's most isolated ethnic groups, shunning contact with outsiders and maintaining a hunter-gatherer lifestyle on their remote island home.

The ocean also has its own sea-people: cultures who live a nomadic life moving from island to island on boats, or building tidal houses on stilts on the shore, such as the Moken peoples of Thailand and Myanmar. In Kerala, India, traditional houseboats known as kettuvallams are still used on the network of lagoons and inlets along the western coast. Meanwhile, nations such as the Maldives and Saudi Arabia are developing concepts for futuristic floating cities as a way to manage climate change and rising sea levels.

Early ocean map
This archaic illustrated map of the Indian Ocean basin is from the Portuguese-made Miller Atlas of 1519. With India at the center, the waters are plied by carracks and galleys.

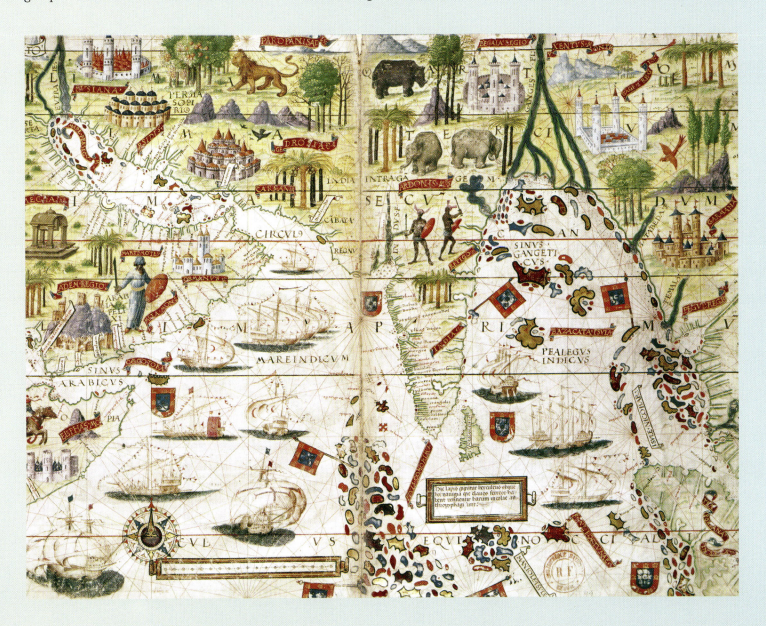

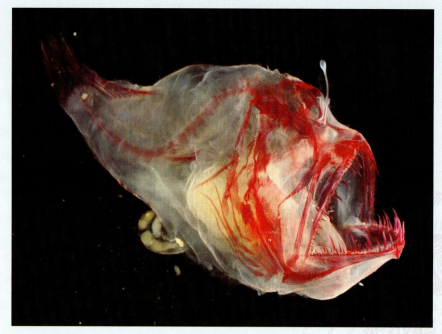

Seadevil specimen
This humpback anglerfish (*Melanocetus johnsonii*)—a species of black seadevil—has been cleaned and stained to identify its bones (red) and cartilage (blue).

Dark Zone

In this part of the deep, dark, and silent depths of the ocean, there are massive waterfalls, rivers without banks, and currents carrying giant waves of sediment.

> **Water** in some areas of the **dark zone** is effectively **stagnant**, which results in **low oxygen levels**

Most of the living space on Earth lies in total, permanent darkness. The dark, or bathyal, zone extends from 3,280 to 13,120 ft (1,000 to 4,000 m) below the ocean surface. The lack of food here is severe, because less than five percent of the material that falls from the surface survives its slow descent to these depths. Temperatures hover a few degrees above freezing and water pressures at the deepest part of the dark zone are 400 times higher than at the surface.

The animals that live here are slow-moving, slow-growing, and long-lived. They often have huge mouths; long teeth; and elastic, balloon-like stomachs to grasp and accommodate prey even larger than themselves. These deep-sea hunters, such as the black swallower (*Chiasmodon niger*), often remain quite small—less than 12 in (30 cm) long—limiting the amount of food they need. Anglerfish (order Lophiiformes) and viperfish (genus *Chauliodus*), which can grow larger, use bioluminescent lures to attract prey. Spookfish, living at the boundary of the twilight and dark zones, have huge eyes and transparent heads to help them identify the silhouettes of prey above.

Vibrant seafloor

The seafloor is crossed by massive canyons where bathyal waters meet the continental slope. These are corridors of food, energy, and oxygen, transported by powerful sediment-laden currents. Sea lilies and sponges cling to the walls, while sea cucumbers and burrowing organisms scour the floors. Contouring the continental slopes, deep underwater currents weave around seamounts and huge sediment drifts; they flow across shallow gateways between ocean basins, and

Fearsome fangs
Migrating between the dark and twilight zones, viperfish such as this have a luminescent lure and fearsome bite. Their long fangs serve to trap prey in their oversize jaws.

tumble down in the largest waterfalls on Earth. One of the largest submarine waterfalls lies beneath the Denmark Strait, east of Greenland, where 180 million cu ft (5 million cubic m) of water cascades down an 11,500-ft (3,500-m) vertical slope every second, all in the silence of the deep ocean.

Such currents also carry nutrients and oxygen to widely dispersed deepwater coral reefs (or mounds) that are unlike their sunlit, tropical counterparts. Most deepwater coral mounds are found in or near pockmarks on the seafloor that mark the natural escape of methane from the sediment. These little-known oases support thriving communities of sponges, worms, mollusks, crustaceans, brittle stars, bryozoans, and sea spiders.

Glowing up
Transparent, fragile, and bioluminescent with gelatinous spikes, comb jellies such as this spotted comb jelly (*Leucothea pulchra*) are unrelated to jellyfish and nonvenomous.

The spotted comb jelly's Latin name means "beautiful sea goddess"

Needlelike teeth form a cage when closed to trap prey

Jaws open more than 120 degrees, helping to swallow extra-large prey

Bioluminescence

The phenomenon of a living organism creating light, known as bioluminescence, is common in the ocean—around 75 percent of deep-sea marine animals are bioluminescent.

> Some **sea cucumbers** detach a **bioluminescent** part of their body to distract **predators** and **escape**

BIOLUMINESCENCE

In the ocean, the light produced by living organisms is usually blue or green, and it happens as the result of a chemical reaction in glandular organs (see below). Bioluminescence is particularly useful to marine life in the deepest, darkest parts of the ocean: below around 650 ft (200 m), there is very little light. Marine life uses bioluminescence for many different reasons, including to communicate or find mates. The female anglerfish (Lophiiformes) has a glowing lure on her head to tempt prey toward her jutting teeth. Others, such as the vampire squid (*Vampyroteuthis infernalis*), produce a bioluminescent mucus when under threat and use it as a distraction while they escape. Some plankton light up when under threat to attract a bigger animal to eat their attacker. Strangely, the velvet belly lanternshark (*Etmopterus spinax*) hides from danger by lighting up. When seen from below, its glowing belly blends in with light from above, making it harder to see.

Light camouflage
Sloane's viperfish (*Chauliodus sloani*) has photophores along the sides of its body to attract prey and to provide camouflage.

Historic observations

Bioluminescence has been noted by humans throughout history. Ancient Greek philosophers Anaximenes (in 500 BCE) and Aristotle (in 350 BCE) described how the ocean lit up when struck with an oar or rod, and in 77 CE, Roman naturalist Pliny the Elder explained how to make medicine from a bioluminescent jellyfish. The phenomenon even impacted World War I naval warfare. In 1918, a German U-boat traveled through a patch of bioluminescent organisms, causing the water to light up like a beacon. The ship was spotted and sunk by Allied forces.

> " One may see fiery sparks when the water is stirred. "
>
> *Hai Nei Shih Chou Chi,* **4th or 5th century BCE**

Blue tide
Jervis Bay, Australia, is known for its glimmering waves, which glow in blues and greens. This bioluminescence is also known as the blue tide and is created by large swathes of algae.

HOW PHOTOPHORES WORK

Bioluminescence produces light through a chemical reaction in organs called photophores. Typically, when the light-producing chemical luciferin and the enzyme luciferase combine with oxygen (and sometimes calcium), the reaction creates light. Marine animals usually make blue light but some have a fluorescent protein that changes the light's wavelength from blue to green.

JELLYFISH PHOTOPHORES

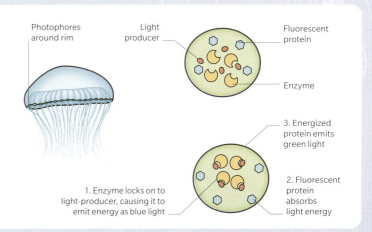

Photophores around rim · Light producer · Fluorescent protein · Enzyme

1. Enzyme locks on to light-producer, causing it to emit energy as blue light
2. Fluorescent protein absorbs light energy
3. Energized protein emits green light

CHIMAERAS 289

Creature of the deep
Chimaeras are known for their bizarre appearance, as exhibited by this spotted ratfish. The first specimen was named *Chimaera monstrosa*, meaning strange or grotesque.

Chimaeras were first described by Swedish biologist Carolus Linnaeus in 1758

Mythical monster
This etching from the late 1700s by Louis Jean Desprez depicts the mythical "chimera"—a three-headed beast that ate humans.

Chimaeras

Chimaeriformes

Also known as ghost sharks, ratfish, rabbit fish, elephant fish, and spookfish, chimaeras are deep-sea fish related to sharks, skates, and rays. They can grow to around 24–80 in (60–200 cm) in length.

Found mainly in the deep ocean worldwide, chimaeras can live at depths of around 1,640–9,800 ft (500–3,000 m). Because there is hardly any light at such great depths, they have large eyes to help them see. Like sharks, they have a cartilaginous skeleton instead of bones, and sensory organs called ampullae of Lorenzini, which detect electrical signals from prey. Some species also have a venomous spine as a form of self-defense.

Despite chimaeras' link with sharks, the two groups diverged from each other around 400 million years ago, and chimaeras show a few key differences to their shark cousins. While sharks have between five and seven gill slits, chimaera have four gills with one opening, covered by a protective flap called the operculum. They also have three pairs of permanent tooth plates that help them crunch and grind hard-shelled prey such as crustaceans and mollusks. These tooth plates, which are fused to the skull, keep growing throughout a chimaera's lifetime.

During mating, males grasp onto females using organs called tentacula. They have one on the forehead and two more in front of the pelvic fins. The female lays the fertilized eggs on the seabed, and they can take between six months and a year to hatch.

EASTERN PACIFIC BLACK GHOST SHARK
Hydrolagus melanophasma
This new species of chimaera was described in 2009 based on samples gathered off the coast of California.

LONG-NOSED CHIMAERA
Harriotta raleighana
This species mainly inhabits the north Atlantic, and can be identified by its distinctive upturned snout.

BENEATH THE WAVES

Sponges

Porifera

There are more than 5,000 species of sponge, most of which live in salt water. These simple animals keep ecosystems clean by filtering water, and have been used by humans for thousands of years.

Sponges have existed for more than 600 million years, and were among the first animals to live on Earth. They are made up of loosely connected cells that work together to filter water through their porous system and extract food particles. Sponges can be made up of different mineral combinations such as collagen, in soft sponges, and hard spicules, seen in glass and calcareous sponges. Scientists have now discovered that some species are carnivorous and others are bioluminescent.

Sponges can vary greatly in size—the largest known specimen, measuring more than 11½ ft (3.5 m) long was found off Hawai'i at a depth of more than 7,000 ft (2,000 m). Sponges are filter feeders that attach to the seabed and cannot move, but if parts of one break off—in a storm or torn off by a predator—they can settle on a new hard surface and begin to grow into a new sponge. Despite this ability to regenerate, sponges have been threatened by overfishing and overharvesting.

Sponges are attached to the mask, giving it a woolly texture

Sponge owl
A 19th-century owl mask from the Solomon Islands. It is made of wood, marine sponge, cane, bark cloth, fiber, and mollusk shells.

SPONGES

Humans have long recognized the benefits of porous, absorbent sponges. Ancient Greeks and Romans used them for cleaning, bathing, and applying cosmetics, and Olympians would use a sponge to daub themselves with oil before competing. Romans may also have used them as padding inside military helmets, and a communal sponge on a stick found in some Roman latrines was likely used as a wipe after using the toilet.

It is not just humans who recognize how useful sponges can be. Dolphins are known to use them to protect their rostrum (nose) when foraging in the seabed.

A perilous profession

People have been diving down to the seabed to collect sponges since around 800 BCE, when ancient Greek divers would harvest them by hand. These sponge fishers could dive down to 100 ft (30 m)—using a piece of marble attached to a rope called a skandalopetra to help them reach the seabed quickly—and stay there for as long as possible to collect sponges. In 1860, the skafandro was introduced—a large, heavy diving suit with a metal helmet and an air pump that went back up to the surface. However, this method put divers at risk of decompression sickness, and many died or became disabled from diving accidents. Although sponge diving is still practiced today, it is banned in several parts of the world.

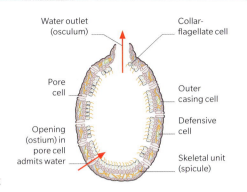

POROUS BODIES

Sponges' porous bodies have bottle-like chambers with loosely connected cells. The lining contains collar-flagellate cells with tiny hairs that beat to keep water flowing through the walls and out of a dedicated opening, trapping food as it passes through.

CROSS SECTION OF TYPICAL SPONGE

Perilous profession
This 1897 lithograph by David Hinds illustrates the labor and risks involved in collecting sponges—including an enormous fish swallowing a fisherman.

VENUS FLOWER BASKET
Euplectella aspergillum

This deep-sea marine sponge has a complex structure, and has inspired architects to explore new, efficient ways of creating 3D-printed structures.

GIANT BARREL SPONGE
Xestospongia muta

Reaching around 5 ft (1.5 m) across, this giant sponge is the largest in the Caribbean. Many animals take shelter in its body.

TUBE SPONGE
Aplysina fistularis

Hawksbill turtles and some reef fish love to eat this sponge that is also sometimes used as a bath sponge.

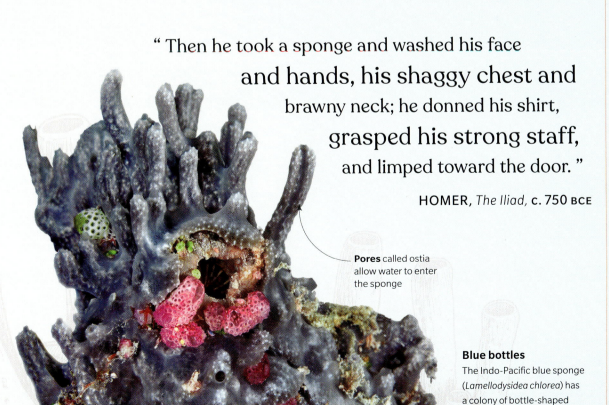

> " Then he took a sponge and washed his face and hands, his shaggy chest and brawny neck; he donned his shirt, grasped his strong staff, and limped toward the door. "
>
> HOMER, *The Iliad*, c. 750 BCE

Pores called ostia allow water to enter the sponge

Blue bottles
The Indo-Pacific blue sponge (*Lamellodysidea chlorea*) has a colony of bottle-shaped chambers that filter the water.

Volcanoes and Seamounts

Volcanoes can throw out lava at temperatures in excess of 1,800°F (1,000°C), and billowing clouds of ash and gas more than 28 miles (45 km) into the stratosphere. Underwater mountains are known as seamounts.

Where fire meets ocean
Red hot lava from Kīlauea Volcano, Hawai'i, cascades over the low cliffs into the ocean. The sudden mixture of seawater with lava and hot rocks can cause explosions as well as acidic steam and gases.

VOLCANOES AND SEAMOUNTS

Vesuvius in Eruption
J. M. W. Turner's evocative painting from around 1820 portrays the volcano that famously destroyed the Roman city of Pompeii in 79 CE.

A volcano is a vent in the Earth's crust that emits hot rock fragments, molten rock, and gases. There are around 1,400 active volcanoes on Earth, many on islands in the ocean. One of the largest volcanic areas is the Pacific Ring of Fire, an underwater belt of volcanoes and tectonic plate boundaries around the Pacific Ocean rim; its volcanoes are often highly explosive with sudden eruptions of rock and ash. This is also where 90 percent of the planet's earthquakes take place.

The Toba supervolcano in Indonesia, which erupted around 74,000 years ago, may have been the largest single eruption on Earth, ejecting as much as 1,300 cu miles (5,300 cu km) of material, the ash cloud reaching the stratosphere (each volcano is unique, and the ash from each location has a distinct chemical signature).

The Krakatoa eruption on Java in 1883 produced the loudest sound ever recorded—it was heard 3,000 miles (4,800 km) away on Rodrigues Island in the Indian Ocean basin. The fine ash from volcanic eruptions can yield vivid sunsets for months or years after the event—the Krakatoa eruption is believed to have been an inspiration for the colorful sky in Edvard Munch's painting *The Scream* (1893). Major eruptions at sea can also cause tsunamis, such as the one caused by the eruption of Santorini around 3,500 years ago.

There are more than 100,000 volcanic seamounts—undersea landforms that rise from the seafloor, but have never broken the ocean surface. Those that do become active volcanoes for a while, but once extinct, they are battered by waves, and slowly subside. The Hawaiian–Emperor seamount chain is a string of sunken volcanoes in the northern Pacific Ocean that once formed above a hot spot. Even Mauna Kea, now 13,796 ft (4,205 m) above sea level, will one day disappear beneath the waves, and join more than 100,000 seamounts below the ocean surface.

Mount Vesuvius's eruption in 1631 killed around 3,000 people

Volcano gods
In folklore, the unpredictable power of volcanoes has evoked many different gods. The word for volcano derives from Vulcan, the Roman god of fire. In Japanese mythology, Konohana-sakuya Hime is the goddess of Mount Fuji and all volcanoes; she is also the blossom-princess and symbol of delicate nature, and an icon of Japanese life.

Hair falls in thin, wispy strands

Statue of Pele
This Polynesian figure from the 17th–18th centuries depicts the Hawaiian fire goddess Pele, a malign force who lives in Kīlauea volcano, Hawai'i. "Pele's hair" is the name given to strands of volcanic glass formed from rapidly cooled lava fountains.

Hydrothermal Vents

There are oases of life in the pitch-black darkness of the muddy seafloor. Here, unusual life forms subsist on the chemicals ejected by hydrothermal vents—fissures created by superheated water that has come into contact with lava in Earth's crust.

Hydrothermal vents eject hot, mineral-rich waters directly onto the ocean floor. They are formed at spreading centers—areas where magma rises up from between tectonic plates—when seawater seeps through fissures and cracks into the seabed crust. When the seawater comes into contact with the lava beneath the surface, it is superheated to 140–867°F (60–464°C) and is ejected back out of the seafloor, enriched with sulfur, iron, copper, zinc, and other minerals that are normally toxic to living creatures. When these waters hit the cold seawater, the dissolved minerals form billowing clouds of particles and cement together into tall, intricate spires that tower above the seabed. The different colored clouds have resulted in the vents being categorized as "black" or "white" smokers.

Stranger than myth

Rich and complex communities surround the vents. These are founded on chemosynthetic bacteria that grow on and within the tissues of the larger organisms. They use the chemical energy from the normally lethal hydrogen sulfide emitted by the vents to produce compounds that are used as food. Organisms unique to vents include Pompeii worms (*Alvinella pompejana*), eelpout fish (Zoarcidae), giant white clams (*Calyptogena magnifica*), and yeti crabs (*Kiwa hirsuta*). Giant tube worms (*Riftia pachyptila*), as thick as a human arm and 10 ft (3 m) long, have red gills protruding from the tips of their white tubes (see p.296).

Marine geologist Robert Ballard discovered hydrothermal vents in 1977, and this led to a new understanding of how organisms survive in extreme underwater conditions. Active vents may be hundreds of miles apart, and they appear to be active for 15–20 years, after which they shut down and the associated life collapses.

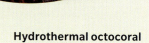

Hydrothermal octocoral
This bright-red species of soft coral thrives near active vents and has a unique ability to detoxify metals.

Seafloor spires
Towering, bright white carbonate spires reach up to 200 ft (60 m) from the seafloor at the Lost City Hydrothermal Field on the Mid-Atlantic Ridge. Unlike black smoker vents, this alkaline vent field discharges methane and hydrogen.

BLACK SMOKERS

At mid-ocean ridges, hot basaltic lava—near 18,000°F (10,000°C)—wells up close to the seafloor and forms new oceanic crust. Seawater circulates through this new crust, leaching minerals in an exchange of chemicals and heat. The superheated water surges upward, gushing out at the seafloor as a hydrothermal vent. Dissolved minerals precipitate out as billowing black clouds.

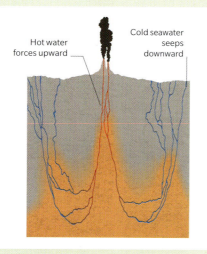

WORKINGS OF A HYDROTHERMAL VENT

> " Wait a minute. What is that? Shimmering water ... coming out right off the top. "
>
> ROBERT BALLARD, on discovering hydrothermal vents, 1977

Giant Tube Worm

Riftia pachyptila

Before giant tube worms and hydrothermal vents were discovered in 1977, scientists were unaware that life could exist without sunlight. In the darkness of the deep sea, these worms subsist on symbiotic bacteria.

Feathery palps extract oxygen from the water

Close relative
Osedax are small deep-sea worms related to the giant tube worm. They survive by eating whale bones, which has resulted in the nickname "zombie worms."

Sometimes known as giant beardworms, these unique animals were discovered in 1977 during a scientific exploration of hydrothermal vents around the Galápagos islands. *Riftia pachyptila* were the first species of tube worm to be discovered, but there are other species living in cold water seeps and at whale carcass fall sites.

Giant tube worms live among hydrothermal vents between 6,200–11,800 ft (1,900–3,600 m) deep, where there is no light and hot water spurts from vents in the seabed (see pp.294–295). This water can reach temperatures of up to 750°F (400°C) and is rich in minerals.

Giant tube worms have no mouth, gut, or anus, so cannot eat. They get nutrients from bacteria in their bodies that take in hydrogen sulfide from the water and turn it into energy. Tube worms are fast-growing organisms, and can reach 3 ft (1 m) long in just 18 months. If hot water stops flowing from the vent, they will die.

Ruby gills
The red plumes are gills that absorb oxygen from the water. They can attract predators, so the worms protect themselves by retracting the gills back into their tubes.

Although giant tube worms are used to a unique, deep-sea environment, scientists have successfully managed to keep them in captivity

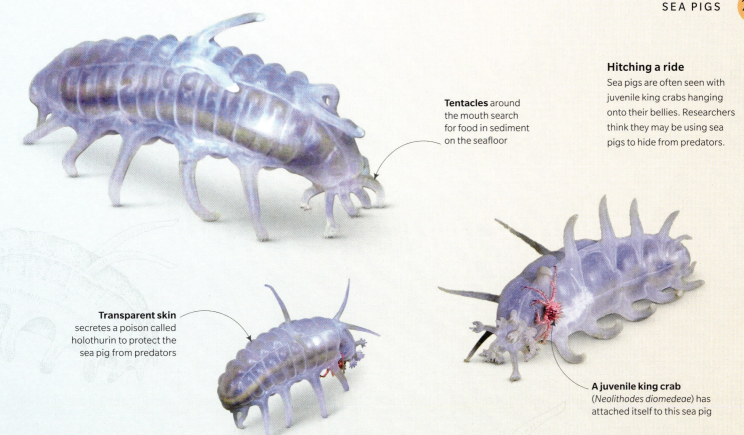

Hitching a ride
Sea pigs are often seen with juvenile king crabs hanging onto their bellies. Researchers think they may be using sea pigs to hide from predators.

Tentacles around the mouth search for food in sediment on the seafloor

Transparent skin secretes a poison called holothurin to protect the sea pig from predators

A juvenile king crab (*Neolithodes diomedeae*) has attached itself to this sea pig

Sea Pigs

Scotoplanes

These sea cucumbers move on tube feet over the top of mud on the seabed looking for food. They consume dead plant and animal matter, and play an important role in keeping the ecosystem healthy.

First described by Swedish zoologist Johan Hjalmar Théel in 1882, sea pigs were among many of the species discovered during the British ship HMS *Challenger*'s around-the-world expedition (1872–1876).

Sea pigs live in waters more than 4,000 ft (1,200 m) deep, down to almost 16,000 ft (5,000 m). They are 4–6 in (10–15 cm) in size, with transparent, water-filled bodies that are designed to cope with the water pressure at deep levels. They are sea cucumbers, related to starfish and sea urchins, but unlike most sea cucumbers, they have pairs of stilt-like feet to help them move through the mud on the seafloor. In addition, their pink color, rotund bodies, and tube feet around their mouths make them look remarkably like pigs. Sea pigs are abundant creatures in their habitat: on some areas of the seabed, they make up 95 percent of all animal life present, by weight. Scientists have discovered that some species are bioluminescent.

Scavenging creatures

There is little food in the deep ocean, so sea pigs are scavengers; they move along the seabed eating particles from dead plants and animals that float down from above. This helps recycle nutrients through the ecosystem. When a whale carcass sinks to the seabed, huge numbers flock to feast on its decaying matter.

Appendages on the sea pig's back are a pair of tube feet. They may help the animal move or sense food

Fluid motion
Sea pigs have between five and seven pairs of hydraulic tube feet, which they inflate with water to move around and find food.

Dumbo octopus
A genus of umbrella octopus, the dumbo octopus (*Grimpoteuthis*) is a deep sea animal that acquired its name from its resemblance to Disney's Dumbo the flying elephant.

Eight webbed tentacles with suckers for steering, foraging, and crawling across the seafloor

Ear-like fins flap to propel the octopus through water

Abyssal and Hadal Zones

No human has ever set foot on the deep-sea floor. Hostile to most life, these areas are home to organisms that have adapted to withstand the huge pressures and pitch blackness.

The abyssal and hadal zones stretch below the sea surface from 13,120–196,680 ft (4,000–6,000 m) and more than 196,680 ft (6,000 m), respectively. Although scientists can survey the deepest seafloor remotely—even from space—using satellite images, few submersibles can reach these depths. These are some of the most unknown and unexplored regions on the planet.

At nearly 6.8 miles (11 km) below the ocean surface, no light penetrates and the hydrostatic pressure is 1,100 atmospheres. This would be deadly for surface dwellers, but deep-sea animals have adapted to survive at such pressures—most lack air spaces such as lungs and swim bladders that could be compressed. These animals include fangtooths (*Anoplogaster brachycera* and *A. cornuta*)—two species of fish with huge mouths, fang-like teeth, and well-honed sensory organs to detect prey—as well as the giant phantom jelly (*Stygiomedusa gigantea*) and sea pigs (*Scotoplanes*) (see p.297). Many of the animals here are seabed scavengers, such as rattails (Macrourinae), hagfish (Myxinidae), and sea cucumbers (Holothuroidea). New organisms are regularly discovered in the sediment. Even below the seafloor in tiny pore spaces, there are bacteria and archaea completely new to science.

The seafloor at these depths comprises vast, flat, smooth abyssal plains, gigantic submarine fans—accumulations of sediment deposited by underwater currents—and deep ocean trenches. The waters are more acidic here than at other levels—the result of higher levels of carbon dioxide released from fallen decomposing organic matter, hydrothermal vent activity, and a lack of photosynthesizing organisms, among other factors. The low carbonate levels dissolve any microscopic shells descending from the surface.

Seafloor explorer
In 2012, film director James Cameron piloted the 24-ft- (7.3-m-) long *Deepsea Challenger* submersible to the bottom of the Challenger Deep in the west Pacific—the deepest-known point on Earth at around 36,000 ft (11,000 m).

World's deepest fish
Snailfish (Liparidae) are scaleless, tadpole-shaped fish found in cold water worldwide. This Mariana snailfish (*Pseudoliparis swirei*), found off the coast of Japan, is the deepest fish ever recorded at a depth of 27,349 ft (8,336 m).

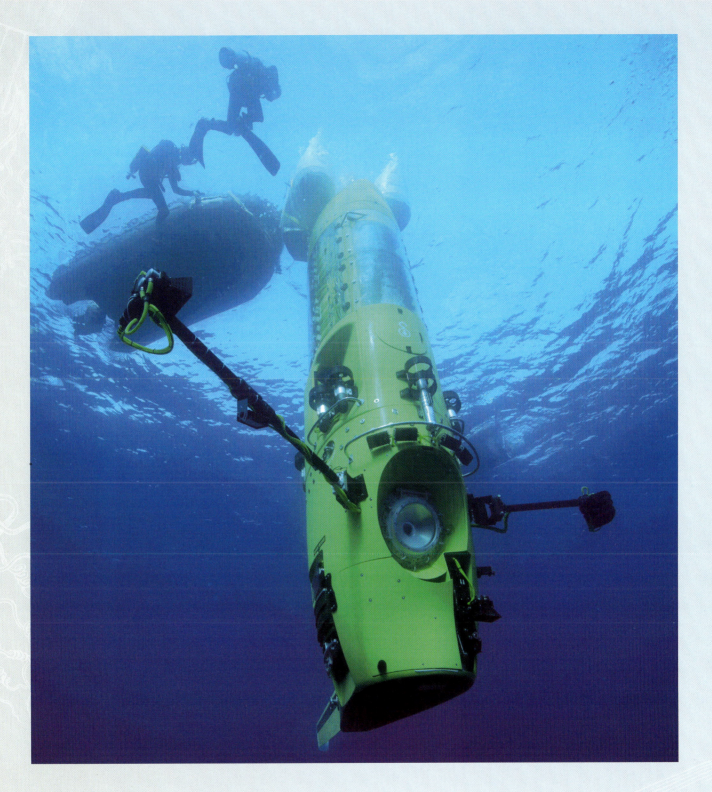

" There are ... places that are special,
 unique and important enough to leave alone—
 and one of these places is the deep. "

HELEN SCALES, *The Brilliant Abyss*, 2021

Trenches

The deepest parts of the world's oceans are long, narrow, steep-sided depressions in the ocean floor, called trenches. Cold and dark, they form unique habitats.

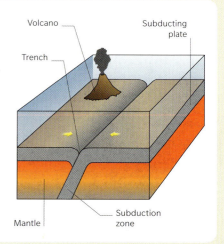

Walking legs are used for mobility

Ocean trenches occur where tectonic plates collide and one is forced under the other. This creates a V-shaped chasm, most of which are more than 18,209 ft (3,000 m) deep. A range of volcanoes often appears around 62 miles (100 km) in advance of where a trench forms, because this is typically where the leading edge of the plate being subducted reaches the mantle and begins to melt (see box below). This process takes millions of years: a tectonic plate is pushed downward at a glacially slow but immensely powerful rate of ½–4 in (1–10 cm) every 12 months, often in a stop-start action that triggers powerful earthquakes and mega-tsunamis.

Exploring trenches

There are around 30 principal oceanic trenches, the majority of which shape the "Ring of Fire" in the Pacific Ocean basin (see pp.292–293). Only a few have been explored. Between 2018 and 2020, however, the Five Deeps Expedition became the first mission to visit the deepest point in each of the world's five ocean basins, using the DSV *Limiting Factor* submersible. This included reaching the bottom of Challenger Deep at the base of the Mariana Trench east of the Philippines. At 35,840 ft (10,924 m) it is the deepest part of the ocean.

Few life forms can live with the near-freezing temperature and crushing pressure of 8 tons per sq in (1.1 metric tons per sq cm) at such depths. However, some creatures have evolved to survive in this habitat, such as the translucent sea pig (see p.297), and the hadal snailfish (*Pseudoliparis amblystomopsis*). Both are gelatinous, with delicate bodies that disintegrate when brought to the surface. Xenophyophores, single-celled organisms measuring up to 8 in (20 cm) across, are widespread.

Trench dwellers
There are many species of squat lobster, flattened crustaceans that typically live on the sea floor. This specimen was photographed in the Mariana Trench.

The **Mariana Trench** was discovered in **1875** during the **Challenger** Expedition

TRENCH FORMATION

Ocean trenches form where tectonic plates in the upper crust, or lithosphere, of Earth's surface collide and the heavier, denser one slides under (subducts) its lighter counterpart. As the heavy plate descends, a deep trench opens up in what is known as the subduction zone. When the solid rock of the descending plate encounters the hot, molten rock of the mantle layer between Earth's crust and its core, it melts and rises back to the surface, usually creating a range of subsea volcanoes.

Explorers of the deep
The mysteries of the ocean depths have always fired the imagination, as this 1950s newspaper illustration recording a 4,000-ft (1,200-m) dive by US inventor Otis Barton in the Pacific shows.

The power of the ocean
Shipwrecks were a popular theme in British Romantic paintings, linking with the movement's sense of awe at nature. Artist J. M. W. Turner (1775–1851) conveyed the tumultuous power of the waves in *The Wreck of a Transport Ship* (c. 1810). This painting shows the ocean at its most terrifying and dangerous: as the seas violently churn, ships are tossed, masts shatter, and humans battle to avoid a watery grave.

Glossary

In this glossary, words in italics have a separate entry.

ABYSSAL Relating to the depths of the ocean. The abyssal plain is a large, flat area of the deep ocean floor, generally at the base of a *continental rise*. The abyssal zone is the area of water between 13,000 and 20,000 ft (4,000 and 6,000 m) deep. See also *Hadal zone*.

AERIAL ROOT A root that grows above the ground, as in many *mangrove* trees.

ALGA (PL. ALGAE) A typically aquatic organism that photosynthesizes but, with the exception of green algae, is not a true plant, lacking roots, stems, and leaves. Examples include seaweeds and many microscopic forms. *Cyanobacteria* are usually now excluded from the definition.

ALLUVIAL FAN A cone-shaped accumulation of sediments formed when fast-flowing streams emerge from a confined channel and then slow down and spread out as they reach an open plain.

ANCHIALINE CAVES Caves that contain water bodies with an underground connection to the ocean and limited exposure to atmospheric oxygen.

ARTHROPODS A major group (phylum) of invertebrate animals with jointed legs, a hard outer skeleton, and a body divided into different segments. It includes *crustaceans* (crabs, shrimp, and their relatives), insects, and spiders.

ATOLL A low, ring-shaped island made of coral, forming the rim of a shallow *lagoon*. The structure results from an accumulation of coral on top of a sunken volcano.

BACKSHORE The part of the shore lying above the normal high-tide line, reached by the sea only under exceptional conditions.

BALEEN A keratinous substance that forms a series of thin plates in the mouths of some whales. It is used to sift food, such as krill, from the water.

BANK In oceanography, a submerged shallow region or plateau of seafloor surrounded by deeper water.

BAR An elongated, narrow, offshore deposit of sediment lying parallel to a coastline. Bars may be permanently submerged, or exposed at low tide. A bar that is always exposed is a *barrier island*. A bar across the mouth of a bay and attached to the coast is a baymouth bar. See also *Barrier island*, *Spit*.

BARRAGE A low, dam-like structure with sluices (gates for water) built across a river or estuary. A tidal barrage is designed to trap water at high tide and obtain hydropower when the water is let out.

BARRIER ISLAND A permanently exposed bar of sand or pebbles lying parallel to a coastline, formed by wind and wave action. A barrier beach is a similar structure, but can be attached to the mainland at one or both ends.

BARRIER REEF A *coral reef* parallel to, but some distance from, a shoreline.

BATHYAL Relating to the area of the sea between the *continental shelf* and the abyssal zone. The bathyal zone lies between 3,300 and 6,500 ft (1,000 and 2,000 m) underwater. See also *Abyssal*.

BATHYMETRY The study of the depth and elevation of water bodies, including the ocean, rivers, and lakes.

BEACH FACE The steeply sloping part of a beach, below a berm. See also *Berm*.

BERM A ridge of sediment high on a beach, left behind by a high tide.

BILLFISH A group of saltwater predatory fish characterized by their large spear-like bills (rostra).

BIOLUMINESCENCE Production of light by living organisms. Sometimes the light is produced by the organism's own cells, at other times by light-emitting bacteria.

BIOME Any large-scale association of organisms, especially one dependent on particular climatic conditions. Mangrove swamp and the abyssal plain are marine biomes.

BIOTA The animals and plants of a particular region or period.

BIVALVE A class of aquatic mollusks with soft bodies encased in hard shells consisting of two halves, or "valves."

BLOOM A large accumulation of *algae* in the surface layer of water. Algal pigments can make water appear discolored.

BRACKISH Describing water that contains more salt than freshwater, but less salt than typical ocean water.

BREAKER ZONE The zone of a beach, or other shoreline, where waves break.

BROOD POUCH A sac or cavity in an animal's body in which eggs develop and hatch.

BUOYANCY The tendency of an object or body to float in a fluid when partially or fully submerged.

BYSSAL THREAD Fibers produced by many species of bivalve mollusks which they use to attach themselves to underwater surfaces.

CAMOUFLAGE Features of an organism that make it difficult to detect visually against its natural background, including color patterns and aspects of body shape.

CARAPACE The protective shield covering the upper surface of a turtle, *crustacean*, or horseshoe crab.

CARTILAGINOUS FISH Fish that have a skeleton wholly or largely composed of cartilage rather than bone.

CARUNCLE A sharp tooth that sea turtle hatchlings use to chip their way out of their shells.

CEPHALOPOD A diverse class of marine mollusks with tentacles, highly developed eyes, and a sac of inky fluid that can be released for defense.

CHROMATOPHORES Pigment-containing cells found in a wide range of animals. Their primary function is to camouflage.

CONTINENTAL CRUST The part of Earth's crust that forms the continents. It is less dense and thicker than *oceanic crust*.

CONTINENTAL RISE A gentle slope connecting the *continental slope* and the deep ocean floor.

CONTINENTAL SHELF The relatively flat and shallow seafloor surrounding a continent, which is geologically part of that continent.

CONTINENTAL SLOPE The sloping seabed that leads from the edge of a *continental shelf* down to the *continental rise*.

CONVERGENT PLATE BOUNDARY An area where two tectonic plates collide and one slides beneath the other. See also *Divergent plate boundary*.

COPEPODS A group of very small *crustaceans* found in nearly every aquatic and semiaquatic habitat. They are one of the most abundant forms of life on Earth.

CORAL BLEACHING Phenomenon in which coral animals lose their symbiotic algae (*zooxanthellae*), usually in response to a stress in the environment, and become bleached of color. See also *zooxanthellae*.

CORAL REEF A rocklike, often ridge-shaped structure of calcium carbonate built by generations of coral animals. See also *Barrier reef*, *Fringing reef*.

GLOSSARY

CORIOLIS EFFECT The tendency for winds and currents moving in a northerly or southerly direction to be deflected because of the effect of Earth's rotation. The deflection is to the right in the northern hemisphere and to the left in the southern hemisphere.

CRUSTACEANS The most diverse and abundant group of *arthropods* in the ocean. It includes crabs, lobsters, shrimp, barnacles, krill, copepods, isopods, and amphipods. Their jointed appendages are variously modified as claws, legs, swimming organs, or filter-feeding devices, depending on the species.

CURRENT A regular, large-scale horizontal or vertical flow of ocean water in one direction; also a smaller flow of water near the shoreline.

CYANOBACTERIA A group of bacteria capable of photosynthesis. Also known as blue-green algae. See also *Alga*.

CYCLONE A large-scale system of winds that circulates around a central region of low atmospheric pressure. See also *Hurricane*.

DARK ZONE Vertical zone of the seabed and water column: in the open ocean this is around 3,300–13,000 ft (1,000–4,000 m), between the *twilight zone* and *abyssal zone*. Virtually no light penetrates this deep.

DELTA An area of low, flat land where a river splits into several branches before flowing into the sea.

DENTICLES Small, toothlike projections, such as the placoid scales of sharks. Their functions include drag (resistance) reduction and protection against predators.

DIATOMS A group of microscopic single-celled organisms featuring an ornamented, box-like protective covering. Most are photosynthesizers and live near the water surface as *plankton*, where they form a crucial part of many marine food chains.

DIEL VERTICAL MIGRATION The daily, synchronized movement of marine organisms between the surface and deep layers of the ocean. This phenomenon is the largest migration of animal life on Earth.

DINOFLAGELLATE Found mostly in the ocean, single-celled organism with two flagella (hairlike appendages).

DIURNAL TIDE A tide cycle characterized by one high tide and one low tide each day.

DIVERGENT PLATE BOUNDARY An area where two tectonic plates are moving away from each other. At these locations, earthquakes are common and magma rises. Under the ocean, these boundaries are where *mid-ocean ridges* form. See also *Convergent plate boundary*.

DOWNWELLING The sinking down of water from the ocean surface. Large-scale downwelling in certain regions gives rise to a deep-ocean current process known as thermohaline circulation. See also *Upwelling*.

DRIFT A large-scale flow of surface water that is broader and slower-moving than a current.

DUNE A mound of sand formed by the action of winds blowing toward coastal land.

ECHINODERM A major group (phylum) of marine invertebrates that includes starfish, sea cucumbers, sea urchins, and sea lilies. Their bodies have radial symmetry, like the spokes of wheel.

ECOTYPE A population of a species that is adapted to a specific environment and forms a distinct group.

EL NIÑO A warming of the equatorial Pacific Ocean that has a profound effect on global climate. See also *La Niña*.

EMERGENT COAST A stretch along the coast that is exposed by a fall in sea level relative to the land.

ENDEMIC Belonging to a particular area.

EROSION The geological process in which the surface of the Earth is worn away by natural forces such as wind or water.

EROSIONAL COAST A coastline where erosion by high-energy waves is the dominant mechanism of change.

ESTUARY The mouth of a large river. Used more broadly, the term includes any bay or inlet where seawater becomes diluted with freshwater.

FETCH The distance of open water over which the wind can blow unobstructed.

FJORD (OR FIORD) An inlet of the sea once occupied by a *glacier*. Fjords are often deep and steep-sided, with a shallower sill at their mouth. See also *Ria*.

FLIPPER A limb used for swimming, for example in whales, seals, penguins, and turtles.

FORAMINIFERA A single-celled amoeba-like organism with a perforated shell.

FORESHORE The part of a shoreline that lies between the average high-water and low-water marks. See also *Tide*.

FRINGING REEF A *coral reef* just offshore, without an intervening *lagoon* or stretch of water. See also *Barrier reef*.

FROND A flattened, leaflike structure found in many seaweeds. Unlike leaves of land plants, seaweed fronds do not contain specialized transportation vessels with which to interchange materials with the rest of the organism.

GASTROPOD A class of *mollusks* that typically have a flattened muscular foot, a shell, and a head bearing sensory tentacles.

GEOGENIC REEF Reefs in which the substratum is not derived from living organisms.

GLACIER A "river" of ice that flows slowly downhill from an ice cap or mountainous region.

GLOBAL CONVEYOR BELT A constantly moving system of deep-ocean circulation driven by differences in temperature and salinity. Also known as thermohaline circulation.

GULF STREAM A warm, swift ocean current flowing northeast off the Atlantic coast of the US from the Gulf of Mexico.

GUYOT A *sea mount* with a flat top.

GYRE A large system of ocean currents circulating round a central point. There are five major ocean gyres.

HADAL ZONE The deepest zone of ocean water, below 20,000 ft (6,000 m), which occurs only in ocean trenches. See also *Abyssal* and *Trench*.

HEADLAND A coastal landform, caused by erosion, that projects out into the sea. See also *Sea arch*.

HOLDFAST The structure that holds organisms such as seaweed to the seafloor. It is not a true root because it does not absorb nutrients.

HURRICANE A storm with high winds of, or in excess of, 74 mph (119 kph).

HYDROTHERMAL VENT A fissure in a volcanically active area of the ocean floor, where superheated water rich in chemicals emerges from the rock.

HYDROZOAN A class of invertebrates within the phylum Cnidaria. They are usually tiny and mostly live in the sea.

ICE AGE Any episode in which Earth's temperatures were much lower than today and ice cover more extensive. The Ice Age refers to a series of such episodes within the last 2 million years, the last ending around 10,000 years ago.

ICEBERG A large fragment of ice broken off, or calved, from the end of a glacier or ice sheet that is in contact with the sea.

ICE SHELF A floating extension of an ice sheet or glacier over the ocean.

INLET An indentation of a shoreline, such as a bay, cove, or estuary. Inlets are usually long and narrow.

INTERTIDAL Between the high- and low-water marks of a shoreline.

ISTHMUS A narrow strip of land with water on either side, connecting two larger areas of land.

LAGOON An enclosed area of coastal water almost cut off from the sea by a barrier; the sea area within an *atoll*.

LA NIÑA A cooling of the central and east-central equatorial Pacific Ocean. See also *El Niño*.

LITHOSPHERE The solid, outer part of Earth, including the crust and upper mantle. The outermost layer of Earth's lithosphere is the *oceanic crust*.

LONGSHORE DRIFT The process by which sediments such as sand and mud are transported along the coastline as a result of sea waves breaking at an oblique angle to the shore.

MANGROVES Salt-tolerant trees that grow in sheltered *intertidal* zones in warmer regions and whose roots are regularly covered by seawater. The term can also refer to the ecosystems and forests that feature such trees.

MARGINAL SEA A division of the ocean that is between the coastal zone and open ocean.

MARINER'S ASTROLABE An ancient navigational tool used to measure the altitude of the sun and stars, allowing sailors to calculate their position at sea.

MESOPELAGIC ZONE See *Twilight zone*.

MID-OCEAN RIDGE An undersea mountain range running along the deep-ocean floor where new oceanic crust is created. See also *Divergent plate boundary*.

MIXED TIDE A tidal cycle that has two high tides and two low tides of different sizes per day.

MOLLUSK Invertebrate, soft-bodied animals of the phylum Mollusca, including clams, octopuses, squid, slugs, and oysters.

MONSOON A seasonal change in the direction of the prevailing winds of a region, causing wet and dry seasons.

MUD FLAT A stretch of very wet soil that is covered at high tide and left uncovered at low tide.

NATAL HOMING The process by which animals migrate away from their geographic area of origin and then return to reproduce. This is undertaken by ocean animals including sea turtles and the Atlantic puffin.

NEAP TIDE The *tide* with the smallest range within an approximately two-week cycle, caused by the gravity of the sun partly canceling out the effect of the moon. See also *Spring tide*.

NEARSHORE The part of the shore affected by waves and tides under normal conditions. It includes the *foreshore* plus an area beyond where the bed is shallow enough to be stirred up by wave action.

NEKTON Any aquatic organism that can swim and move independently of water currents. See also *Plankton*.

NINTH WAVE A nautical phrase to describe a colossal wave that comes after a series of waves increasing in size.

OCEAN BASIN A region of low-lying *oceanic crust* within which deep water is contained, and usually surrounded by land or shallow seas.

OCEANIC CRUST Earth's crust under the deep oceans, which is thinner and denser than the crust of the continents. See also *Continental crust*.

OCEANIC GATEWAY A passage where water masses from different *oceanic basins* meet.

OUTGASSING The emission of a gas previously trapped inside a material. Ocean outgassing occurs when ocean water begins to warm.

PACK ICE Floating *sea ice* that is not attached to land. It may be fused together in a single mass or consist of separate ice floes.

PANCAKE ICE A stage in the formation of *sea ice* in turbulent waters consisting of separate floating plates. Pancake ice plates have raised, rounded edges as a result of colliding with other plates.

PECTORAL FIN Either of the front pair of fins in most fish and marine mammals, mainly used for steering but sometimes for propulsion. See also *Pelvic fin*.

PELVIC FIN Either of the pair of fins located farther back than the *pectoral fins* in most fish.

PHOTIC ZONE See *Sunlit zone*.

PHOTOPHORE A light-emitting organ in certain fish and other bioluminescent animals. See also *Bioluminescence*.

PHOTOSYNTHESIS The chemical process by which plants and some other organisms transform light energy into chemical energy.

PHYTOPLANKTON Small, freely floating, plantlike organisms on which sea creatures feed. Also known as microalgae.

PLANKTON Any organism that does not have the size, energy, or capability to swim against tides or ocean currents. See also *Nekton*.

POD A group of marine mammals that swim together.

POLYP An invertebrate with a cylindrical body and tentacles around the mouth.

REEF A ridge of relatively stable material, usually rock or coral, lying near the surface of the sea. See also *Coral reef*.

RIA A coastal inlet formed by the partial submergence of a river valley.

ROGUE WAVE A huge and unpredictable wave that is likely to cause damage.

ROOKERY A breeding colony of penguins, and sometimes of other seabirds or seals.

SALINITY The degree of saltiness of seawater.

SALP A transparent marine animal that has a barrel-shaped body with an opening at each end.

SALT MARSH An ecosystem that may develop on the upper *intertidal* zone of sheltered, muddy coastlines, mainly in cooler parts of the world, and consisting of a community of small, salt-tolerant land plants. In the tropics, *mangrove* forests generally grow in similar situations.

SALT PAN An undrained shallow depression in which saltwater evaporates to leave a deposit of salt. Also known as a salt flat.

SEA Sometimes used as another word for ocean, but can also refer to a relatively small, often shallower region of an ocean partly marked off by land.

SEA ARCH A natural arch on a rocky shoreline, usually created by two *sea caves* on either side of a *headland* eroding into each other.

SEA CAVE A cave created at the foot of a cliff by wave action.

SEA ICE Ice formed directly on the sea, as distinct from *icebergs* and *ice shelves*, which originally formed on land before entering the sea. See also *Pack ice*, *Pancake ice*.

SEAGRASS One of many flowering grasslike plants that have adapted to live under saline water.

SEA MOUNT An undersea mountain, usually formed by volcanic activity. See also *Guyot*.

GLOSSARY

SEAWEEDS Any of several groups of relatively large marine, plantlike organisms traditionally classified as multicellular algae. One group, the green seaweeds, is closely related to land plants.

SEDIMENT Particles carried in water that are capable of settling under the force of gravity, or deposits formed by such particles, including sand, silt, or mud.

SEMIDIURNAL TIDE A tidal cycle that has two high tides and two low tides of roughly the same size per day.

SESSILE ANIMAL An animal that is permanently anchored to a substrate and cannot move freely.

SHOAL A large group of fish swimming together. Or a submerged ridge, usually of sand, that forms in rivers and oceans.

SIRENIANS Order of large, aquatic, plant-eating mammals that have forelimbs resembling paddles and no hind limbs. Includes manatees and dugongs.

SPAWN A release or deposit of eggs by fish, amphibians, or mollusks.

SPIT A peninsula of sand, shingle, or both created by *longshore drift*, usually at a point where the shoreline changes direction. See also *Bar*, *Barrier island*, *Tombolo*.

SPONGES A phylum of simple-bodied, slow-moving animals that live by straining food particles from the water.

SPREADING RIDGE A *mid-ocean ridge* where Earth's tectonic plates move apart, creating new oceanic crust through rapid volcanic activity.

SPRING TIDE The highest high tide and lowest low tide within an approximately two-week cycle, caused by the sun and the moon being in positions in which their gravitational effects combine to most effect. See also *Tide*, *Neap tide*.

STRAIT A narrow passage of water that connects two larger bodies of water.

SUBDUCTION The process of two tectonic plates colliding and one descending below the other.

SUBDUCTION ZONE A location on Earth where two tectonic plates collide and one dives beneath the other.

SUBMERGENT COAST A coastline that has been flooded due to a rise in sea level relative to the land.

SUNLIT ZONE The topmost layer of ocean water, where enough light penetrates for photosynthesis to occur. Also called the photic zone, in clear waters it can extend from the surface down to 660 ft (200 m). See also *Dark zone*, *Twilight zone*.

SUPERCELL THUNDERSTORM A dangerous type of thunderstorm with a deep rotating updraft called a mesocyclone.

SURFACE CURRENT A current at the ocean surface, usually driven by wind.

SWELL A series of waves that propagate over long distances, influenced primarily by gravity.

TIDAL BORE A single large wave created when an incoming tide moves up a narrowing channel, such as an *estuary*.

TIDAL BULGE The highest point of water caused by the moon's gravitational pull.

TIDAL FLAT A flat, muddy area covered at high tide; characteristic of sheltered areas such as estuaries.

TIDAL ISLAND A raised area of land that is connected to the mainland at low tide but separated at high tide.

TIDAL RANGE The difference in height between high and low tide at any given location.

TIDE The regular variation in the height of the ocean surface at a given point, caused by the gravitational attraction of the moon and sun combined with Earth's rotation. In coastal regions, it results in horizontal movements of water.

TIED ISLAND An island connected to the mainland only by a *tombolo*.

TOMBOLO A narrow area of sand that joins an island to the mainland.

TRADE WINDS Prevailing winds that blow east to west, circling the Earth near the equator. They are also known as easterlies; westerlies blow from west to east in the mid-latitudes.

TRENCH A large-scale, naturally forming chasm on the ocean floor where the ocean is deepest and one tectonic plate is forced under another. The deepest is the Mariana Trench in the Pacific Ocean.

TROUGH A long, wide, and deep depression in the seafloor that is characteristically shorter, shallower, and narrower than an oceanic *trench*.

TSUNAMI A large, fast-moving ocean wave generated by an undersea earthquake or a large coastal landslip.

TUBE FEET Fleshy, tubular projections on the bodies of most *echinoderms*, used to move, feed, and breathe.

TUSK An enlarged tooth of a mammal such as a walrus or narwhal.

TWILIGHT ZONE In the open ocean this layer is between approximately 660 ft and 3,300 ft (200 and 1,000 m) deep, into which some light penetrates, but not enough to support *photosynthesis*. Also known as the mesopelagic zone.

TYPHOON A violent tropical storm with very strong rotating winds.

UPWELLING The rising of deep-ocean water to the surface. It can be caused by wind blowing parallel to a coastline or by an underwater obstruction such as a sea mount interrupting a deep sea current. Upwelled water often enriches the surface of the ocean with nutrients. See also *Downwelling*.

WAVE A motion or disturbance that transfers energy. The water in a wave crossing the open ocean does not move significantly except up and down as the wave passes. The high point of a wave is its crest and the low point its trough.

WETLAND An area of land that is permanently or seasonally saturated in water.

WHIRLPOOL A powerful eddy or vortex formed at the sea's surface, often caused when two separate tidal currents meet.

ZOOPLANKTON Any animals or animallike protists (unicellular organisms) that are part of the *plankton*.

ZOOXANTHELLAE A group of yellow-green or yellow-brown microalgae that inhabit other organisms such as marine invertebrates.

Index

Page numbers in **bold** refer to main entries.

A

A-76 iceberg 184
abalones 53, 54–5, 119, 121, 261
Aboriginal Australians 55, 86, 95, 101, 207, 210, 210
abyssal zone 13, 32–3, 268, **298–9**
Acanthaceae (acanthus) 128
Acanthaster planci (crown-of-thorns starfish) 235
Actiniaria spp. (anemone) 57
Aelianus, Claudius, *On the Nature of Animals* 209
Aethia pygmaea (whiskered auklet) 41
African mythology 233, 233, 266
Agulhas Current 150
Ainu people 257
Aipysurus duboisii (Dubois' sea snake) 264
air currents 143, 145, 162
Aivazovsky, Ivan, *The Ninth Wave* 150
Alaska 43, 111, 180
albatrosses 140, **144–5**, 185, 188
Alcidae (auk) 188
Aldabrachelys gigantea (Aldabra giant tortoise) 134, 282
Aldrovandi, Ulisse 223
Aleut people 121
Alexander, Meena 282
Alexander the Great 197
algae **36–7**, 67, 78, 170, 237
    in plankton 197, 199
    seaweeds 118, 202
    zooxanthellae 161, 206, 207, 211, 270
al-Hariri of Basra 281
Aliger gigas (queen conch) 55
Alvinella pompejana (Pompeii worm) 294
Amblyrhynchus spp. (iguanas) 47
A. cristatus (marine iguana) 136–7, 137
America's Cup 140
ammonites 17, 62, 63, 63, 81
Ammophila spp. (marram grass) **99**
A. arenaria (European marram grass) 99, 101
A. breviligulata (North American marram grass) 99
Ammospiza caudacuta (saltmarsh sparrow) 123

Amphioctopus marginatus (veined octopus) 255, 257
amphipods 69
ampullae of Lorenzini 243, 263, 289
Anarchias leucurus (finespot moray) 214
Anarrhichthys ocellatus (wolf-eel) 237
Anaximenes 287
Andaman Island 282, 283
anemones 57, 70, 161, 205, 216, 237
angelfish 92–3
anglerfish 159, 159, 284, 285, 287
Anglo Saxon Riddles 61
Anguilliformes **214–15**
Anguilla anguilla (European eel) 159
animalcules (little animals) 35
animals 29, **34–5**
    see also birds; fish; invertebrates; mammals; reptiles
Anoplogaster spp.
    A. brachycera (fangtooth) 298
    A. cornuta (fangtooth) 298
Antarctica 17, 25, 189, 192
    ice 170–1, 173
    oceans 184–5, 258
Antarctic Circumpolar Current 14, 25, 150, 184
Antarctic Convergence 184
Antarctic Ocean (Southern Ocean) 12, 24, **184–5**, 186–7
    ice 170–71, 173
    storms 150
antlers 173, 213
Aplysina stularis (tube sponge) 291
Aptenodytes forsteri (emperor penguin) 185, 188–9, 189, 190
Arabian Peninsula 165, 283
Arabian Sea 15, 280
Aral Sea 14
Arbacia lixula (black sea urchin) 67
archaea 17, 298
arches **44–7**, 171
Architectonicidae (sundial) 54
Architeuthis dux (white giant squid) 34
Arcos poecilophthalmos (Galápagos red clingfish) 137
Arctic Ocean 12, 170, 172, **180–81**, 182, 239
    ice 170–71, 171, 173
    narwhals 178
    polar bears 175, 176
    whales 116, 117, 225
Ardeidae (waders) 108, **126–7**
Arecaceae (palm tree) 96, 128
Arenaria interpres (ruddy turnstones) 73
Arenicola spp. (lugworms) 85, **98**
A. defodiens (black lugworm) 98

A. marina (blow lugworm) 98
Argonauta hians (brown paper nautilus) 257
Aristotle 36, 66, 68, 71, 196, 217, 287
arribadas (mass nestings) 92–3
Artemisia halodendron 124–5
arthropods 35, 68, 69, 123
Ascension Island 233
astrolabes 282
Atlantic Ocean 12, 24, 152–3, **230–33**
    animals 42, 43, 55, 106, 123, 198
    hydrothermal vents 295
    Namibian coastline 104
    Sargasso Sea 158–9, 214
    storms 143, 146, 150
Atlantic Ocean islands 132, 233
    Azores 132, 233, 239
    Bermuda 58–9, 158, 233
    Caribbean islands 55, 233, 239, 244, 266, 279
    Faroe Islands 47, 233
    Newfoundland 150, 165, 171, 173, 233
    see also Great Britain; Iceland; Ireland
Atlantis 239, 239
atmosphere 16, 20, 140
atolls 133, 145, 206, 282
Audubon, John James 79, 126, 167
auklets 41
auks 115, **188–9**
Australia
    animals in 130, 209, 221
    beaches 31, 84, 85, 86–7, 286
    Great Barrier Reef 205, 207, **210–11**, 235
    Indigenous peoples 55, 86, 95, 101, 207, 210, 210
    islands 101, 134, 135, 188
Australian pelican **94–5**
avalanches 101
avocets 127
Azores 132, 233, 239
Aztec Empire 146

B

Babylonia lutosa (mud ivory whelk) 55
Bach, Richard, *Jonathan Livingston Seagull* 156
Bacillariophyceae (diatom) 199

bacteria 17, 123, 196, 250, 298
    cyanobacteria 35, 197
    in hydrothermal vents 294, 270
    phytoplankton 23, 173, 181, 197, 198, 199
Bahamas 85, 128, 158
Balaena mysticetus (bowhead whale) 173
Balaenoptera spp.
    B. acutorostrata (common minke whale) 226
    B. brydei (Bryde's whale) 225
    B. musculus (blue whale) 12, 225, 226
    B. omurai (Omura's whale) 225–6
baleen whales 225, 226
Ballard, Robert 294
Baltic Sea 82
Bangladesh 110, 128
Barbut, James, *The Genera Vermium* 273
barnacles 57, 69, 71 237
barracudas 33
barrier islands 76, 77, 133
barrier reefs 206
    Great Barrier Reef 205, 207, **210–11**, 235
barriers, natural 25, 76, 111
Barton, Otis 301
basalt 12
basins **12–13**
    see Arctic Ocean; Atlantic Ocean; Indian Ocean; Pacific Ocean; Southern Ocean
basket stars 250
Basque region 274
bass 118–19
batfish 280
bathyal (dark) zone 284
bathymetry 248
Bathymodiolus thermophilus (giant mussel) 270
Bathyteuthida (squid) **278–9**
Baxter, William 75
bays 14, 84, 122
    formation of 30, 45, 47, 76
beach grass 101
beaches 31, 45, 98
    Chesil Beach, Dorset **80–81**
    for reproduction 89, 282
    rocky **56–7, 72–3**, 127
    sandy **84–7**
bears, polar 173, 173, **174–7**, 181
Beebe, William 33
belemnites 62, 81
Belgium 176
beluga whales **116–17**, 173, 178, 225

INDEX

Benguela Current 104, 105, 165
Benin 76
Bering Sea 14, 175
Bering Strait 173
Bermuda 58–9, 158, 233
Bermuda Triangle 159
Biard, François-Auguste 180
billfish 266–7, 275
bioluminescence 250, 251, 284, **286–7**, 290, 297
    jellyfish 216–17
    squid 32, 279
birds
    albatrosses 140, **144–5**, 185, 188
    cormorants 46, **48–9**, 94
    flamingos **78–9**, 105
    Galápagos Islands 137
    gulls 42, 127, **156–7**, 167
    pelicans **94–5**, 123, 123
    penguins 115, 137, 185, **188–91**, 192
    puffins 41, **42–3**
    terns 85, 101, 108, 127, 157, **166–7**
    waders (birds) 110, **126–7**
    wetlands 124–25
Birgus latro (coconut crabs) 70, 96
bivalves 53, 127, **270–3**
Blackbeard 87, 233
Black Sea 15, 68
blacksmith fish 118–19
black smokers 136, 294, 295
black swallowers 32, 284
blennies 65, 237
blooms 37, 173, 197, 199, 219
blubber 107, 113, 117, 121, 156, 183
boats 157, 210, 260, 283
    see also ships
Bodianus diplotaenia (Mexican hogfish) 92–3
Bonny, Annie 133
bony fish 210, 252
Bora Bora 77, 196
Botticelli, Sandro, *The Birth of Venus* 271
box jellyfish 216
brackish water 110
bradyseism 239
Brazil 86, 100, 249
breaching 226, 228–9
Brendan the Navigator 230
Bridges, Robert 75
bristlemouths 251
brittle stars 234, 237, 285
Brontë, Anne 85
bryozoans 63, 285
bubble-net feeding 227
Burne-Jones, Edward, *Pelican in Her Piety* 95
Buteo galapagoensis (Galápagos hawk) 137

Butorides sundevalli (lava herons) 137
byssal threads 57
by-the-wind sailors **160–61**

C

Calidris
    C. alpina (dunlin) 123
    C. ferruginea (curlew sandpiper) 127
    C. pygmaea (spoon-billed sandpiper) 124
Callistoctopus macropus (white-spotted octopus) 254
calving 171, 184
Calyptogena magnifica (giant white clam) 294
Cambodia 96
Cambrian period 16, 35
Cameron, James 299
Cameroon 96
Canada 45, 170, 230
    animals in 117, 178, 178–9
    Arctic Ocean 180, 181
    First Nations 53, 55
    fjords 115
    Newfoundland 150, 165, 171, 173, 233
Canaries 233
Canary Current 158
canyons 25, 284
Cape Verde 233
carapaces 68, 88, 88
Carboniferous period 16
Carcharhinus spp.
    C. galapagensis (Galápagos shark) 47
    C. leucas (bull shark) 108
    C. limbatus (blacktip shark) 241
Carcharias taurus (sand tiger shark) 276–7
Carcharocles megalodon (Megalodon) 34
Carcharodon carcharias (great white shark) 159, 240
Carcinus maenas (shore crab) 68
Caretta caretta (loggerhead turtle) 90, 159
Caribbean islands 55, 233, 239, 244, 266, 279
Caribbean Sea 14
Carpilius maculatus (alakuma crab) 71
Carroll, Lewis
    Alice in Wonderland 78
    "The Walrus and the Carpenter" 183, 183
Carta marina et description septemtrionalium terrarum (Marine map and description of the Northern lands) 8–9
cartilage 241

Caspian Sea 14
Cassell's Natural History 136
Cassiopea spp. (upside-down jellyfish) 217, 219
casts 98
caves, sea 45, 47, **50–51**
Cayman Islands 244
cells (air movement) 140–42
Celtic mythology 165
cephalofoils 263
cephalopods
    cuttlefish 278
    nautiluses 257
    octopuses **254–7**, 278, 298
    squid 32, 34, 251, **278–9**, 287
Cephalorhynchus heavisidii (Heaviside's dolphin) 220
Cepphus columba (auks) 115
cerata 52, 53
Cetacea (whales) **224–7**, 228–9
Cetorhinus maximus (basking shark) 242
Chaetodipterus faber (Atlantic spade fish) 276–7
chalk 45
Challen, Gordon Leslie 250
Challenger Deep 300
channels 31, 47, 108, 109, 111, 162
Charadriidae (waders) 85, **126–7**
    Charadrius hiaticula (ringed plover) 73, 101
Chauliodus spp. (viperfish) 33, 284
    C. sloani (Sloane's viperfish) 287
Cheilopogon melanurus (Atlantic flying fish) 33
Cheloniidae (sea turtles) 85, **88–91**
    Chelonia mydas (green sea turtle) 88, 133, 159
Chelonoidis niger (Galápagos giant tortoise) 136
Chesil Beach, Dorset **80–81**
Chiasmodon niger (black swallower) 32, 284
Chile 35, 115, 165, 249
chimaeras **288–9**
Chimaeriformes spp. (chimaera) **288–9**
Chimaera monstrosa (ratfish) 288
China 29, 49, 96, 110, 124–5, 235, 283
    mythology 91, 261
Chironex fleckeri (Australian box jellyfish) 219
chitin 68, 71
Chlidonias spp. (marsh tern) 167
Christianity 227, 272
Christie, Agatha 83
Christmas Island 282
Chroicocephalus novaehollandiae (silver gull) 156
chromatophores 208, 257, 279

Chromis punctipinnis (blacksmith fish) 118–19
Chrysaora spp.
    C. achlyos (black sea nettle) 216–17
    C. fuscescens (Pacific sea nettle) 219
Chumash people 221
Cirripedia (barnacle) 57
cities 110, **238–9**
clams 85, 270, 270, 272, 294
Clark, Eugenie 243
cliffs 30, **40–41**, 56
climate **26–7**
    see also weather
climate change 77, 86, 128, 147, 173, 211
    threats to animals 176, 181, 182, 243
clouds 164–5
cnidaria 161
    anemones 57, 70, 161, 205, 216, 237
    by-the-wind sailors **160–61**
    see also corals; jellyfish
coastal landscapes 30–31
    arches **44–7**, 171
    cliffs 30, **40–41**, 56
    coastal deserts 104–5
    deltas 31, **108–11**, 122, 173
    dunes 75, 99, **100–101**, 104
    estuaries 29, **108–11**, 122, 127, 130
    fjords 30, **114–15**
    mangroves **128–9**, 130
    mudflats 108, **122–3**
    rock pools **64–5**, 237
    salt marshes 108, **122–3**
    sea caves 45, 47, **50–51**
    sea stacks 30, **44–7**
    tied islands 81, **82–3**
    see also beaches; lagoons
cockles 123, 270
coconut palms **96–7**, 133
Cocos nucifera (coconut palm) **96–7**, 133
coelacanth fish 282
Coelacanthiformes 35, 282
cold currents 22
Coleridge, Samuel Taylor, *The Rime of the Ancient Mariner* 145
collagen 290
Columbus, Christopher 36, 113, 159, 230
columns (sea stacks) 30, **44–7**
comb jellies 199, 284
Combretaceae (mangrove) 128
Comoros 282
conchs 54, 55
concordant coastlines 30

INDEX

Condé, Maryse, *Crossing the Mangrove* 128
Coniopteris spp. 63
Conolophus subcristatus (land iguana) 137
conservation 86, 91, 145
continental drift 16–17, 63
Conus pennaceus (feathered cone snail) 55
convergent plate boundaries 18, 19
conveyor belt 24–5
Cooper, James Fenimore 133
copepods 197, 199, 251
coral atolls 206, 282
coral bleaching 211
Corallinales spp. 202
coralline algae 202
coral reefs 133, 197, **204–7**, **210–11**, 235, 282
corals 85, 87, 161, 216, 285
   cold-water 237
   near hydrothermal vents 294
   in mangrove forests 129
Coral Sea 14
Coriolis effect 23, 140, 142, 147, 230
cormorants 46, **48–9**, 94
corrasion 41
Coryphaena hippurus (dolphinfish) 159
Cotier peoples 283
Cotylorhiza tuberculata (fried egg jellyfish) 217
Cousteau, Fabien 239
Cousteau, Jacques 35, 197, 239
coves 30
crabs **68–71**, 85, 85, 96, 159, 294, 297
   Christmas Island red crabs 282, 282
   hermit crabs 65
Crambe maritima (sea kale) 73
crayfish 68
Creagrus furcatus (swallow-tailed gull) 157
creeks 108, 109
Cretaceous period 17
Crinoidea (feather star) 235
crinoids 81
crocodiles 108, **130–31**
Crocodylus porosus (saltwater crocodile) **130–31**
Crossaster papposus (common sun star) 237
crust 17, 18, 294
Crustacea (crustaceans) **68–71**, 251, 272, 300
   barnacles 57, 69, 71 237
   copepods 197, 199, 251
   isopods 110
   krill 69, 71, 173, 186–7, 199

Crustacea (crustaceans) continued
   lobsters 68–9, 70, 71, 300
   shrimp 68, 71, 78, 123, 159, 282
   see also crabs
Ctenophora spp. (comb jelly) 199
curlews 126
currents 140, 143, 162, 165
   in Atlantic Ocean 104, 158, 230
   deep-water **24–5**, 284, 298
   formation of garbage patches 260
   in Pacific Ocean 137, 150
   in Southern Ocean 150, 184
   surface **22–3**, 26, 81, 82, 108
cuttlefish 278
Cuvier, Georges
   The Animal Kingdom 252
   Discours sur les Révolutions du Globe 63
Cyanea capillata (lion's mane jellyfish) 199
cyanobacteria 35, 197
cyclones 143, **146–7**, 260, 280
Cymodoceaceae (grasses) **102–3**
Cyphoma gibbosum (flamingo tongue snail) 54
Cyprus 47

D

dark zone 32–33, **284–5**
Darwin, Charles 206, 257
   on HMS *Beagle* 35, 36, 119, 137, 137, 206
Dasyatis pastinaca (common stingray) 245
decapods 68
Deepsea Challenger submersible 299
deep-sea life 37, 284–5
   see also ocean zones
deep-water currents **24–25**
Defoe, Daniel, *Robinson Crusoe* 134
de la Beche, Henry 63
Delphinapterus leucas (beluga whale) **116–17**, 173, 178
Delphinidae **220–21**
   Delphinus delphis (common dolphin) 221
deltas 31, **108–11**, 122, 173
Demarle, Alphonse 235
Denmark 62, 181
   Faroe Islands 47, 233
   see also Greenland
Denmark Strait 248, 285
denticles 240
deposition (geology) 25, 31, 77
Dermochelyidae spp. (sea turtle) **88–91**

Dermochelyidae spp. (sea turtle) continued
   Dermochelys coriacea (leatherback sea turtle) 88, 89, 90
desert islands 96, **132–5**
deserts **104–5**, 165, 170
Desmonema annasethe (jellyfish) 218
Desprez, Louis Jean 289
Devonian period 16
diatoms 36, 199
Dickens, Charles, *A Tale of Two Cities* 73
diel vertical migration (DVM) 33
dinoflagellates 199
Diomedeidae (albatross) **144–5**
   Diomedea irrorata (waved albatross) 145
discordant coastlines 30, 45
distributaries 31
diurnal tides 29
divergent plate boundaries 18
diving, underwater 290, 291
dodos 282
doldrums 142
dolphinfish 159
dolphins **220–21**, 291
   orcas 35, 60, 192, 220, **222–3**
doomsday fish 253
Doré, Gustave 221
Dosidicus gigas (Humboldt squid) 278
dowitchers 126
downwelling 23
Drake Passage 14, 25
dread (bird flight) 43
Dreamtime stories 210–11
Dryococelus australis (Lord Howe Island stick insect) 135
DSV Limiting Factor submersible 300
Dugong dugon (dugong) **112–13**
dunes 75, 99, **100–101**, 104
dunlins 123, 123
Durvillaea willana (bull kelp) 46

E

Earle, Sylvia 36, 159
earthquakes 18, 19, 253, 293
   and tsunamis 154, 155, 260, 300
Easter Island 167
Echinodermata spp. (echinoderm) **234–5**
echinoderms **234–5**
   basket stars 250
   brittle stars 234, 237, 285
   sand dollars 67
   sea cucumbers 235, 235, 284, 286, **297**, 298
   sea lilies 235, 284

echinoderms continued
   sea urchins 35, 57, 65, **66–7**, 235, 237
   starfish 67, 119, 235, 237
Echinoidea spp. (sea urchins) 57, **66–7**
Echinothrix calamaris (banded sea urchin) 235
Echinus esculentus (edible sea urchin) 67
echolocation 117, 223, 225
ecotypes 222
Ecuador 136
eddies 108
eelpout fish 294
eels 102, 123, 159, **214–15**, 237, 252
Egede, Hans 265
egrets 123, 127
Egretta tricolor (tricolored heron) 126
Egypt
   ancient 143, 202, 209, 239, 239
   modern 51, 110
Elasmobranchii (sharks) **240–43**
electrical signals 243, 263, 289
elements, of seawater 20
elephant fish 289
elephants 282
El Niño 26
emergent coastlines 30
England 87, 149
   Dorset 41, 62, 63, **80–81**
   islands 60, 82, 83
   mythology 81, 165
Enhydra lutris (sea otter) **120–21**
Ensis leei (razor clam) 272
Enteroctopus dofleini (giant Pacific octopus) 255, 255
Eocene period 17
Epinephelus lanceolatus (Queensland grouper) 210
Equator 22, 140–42, 260
Eretmochelys imbricata (hawksbill sea turtle) 90, 282
Eriksson, Leif 230
erosion 25, 31
   of coastal landscapes 41, 45–6, 50, 56
   fossil exposure 63
   of islands 82
escarpments (cliffs) 30, **40–41**, 56
Estonia 82
estuaries 29, **108–11**, 122, 127, 130
Etmopterus spp.
   E. perryi (dwarf lantern shark) 240
   E. spinax (velvet belly lanternshark) 287
Etruscans 209

Eudyptes spp.
  E. chrysocome (southern rockhopper penguin) 191
  E. chrysolophus (macaroni penguin) 189, 191
  E. pachyrhynchus (Fjordland penguin) 115, 191
Eudyptula novaehollandiae (Australian little penguin) 188
Euphausiacea (krill) 69, 199
Euphausia superba (Antarctic krill) 71, 186–7
Euphorbia paralias (sea spurge) 101, 101
euphotic zone 196
Euplectella aspergillum (Venus flower basket) 291
evolution, ocean **16–19**, 37
exoskeletons 68, 70, 71
explorers 145, 258–9, 282–3
  deep-sea 300
  see also submersibles
  early maps 8–9, 15, 110, 248, 259, 283
  polar 45, 171, 181, 184, 189
extinction 34, 35, 235, 240, 275
  birds 167, 188, 282
  sea cows 113
  turtles 88
  see also fossils
extremophiles 78

F

falcons 41
Faroe Islands 47, 233
feather stars 234, 235
Ferrel cells 142–3
fetches 149, 150
Fiji 204, 241, 257, 261
filter feeding 225, 272
fiords *see* fjords
firebirds 78
fish 35
  billfish **266–7**, 275
  chimaeras **288–9**
  deep-sea 285, 298
  eels 102, 123, 159, **214–15**, 237, 252
  Galápagos Islands 137
  giant oarfish **252–3**
  in kelp forests 118–19
  leafy sea dragons **212–13**
  rays 33, 196, **244–5**, 246–7
  in Sargasso Sea 158, 159, 159
  seahorses **208–9**, 212–13
  sea serpents 253, 265–6, 265, 266
  symbiosis 92–3

fish *continued*
  tuna **274–5**
  see sharks
fishing industry 55, 81, 181, 210, 258, 272
  overfishing 245, 275
Five Deeps Expedition 300
fjords 30, **114–15**
flamingos **78–9**, 105
fleas 69
flooding 78, 110–11
floor *see* ocean floor
flounders 249
flying fish 33
fogs 105, **164–5**
folklore *see* mythology
foraminifera 85
forests, kelp 66, 67, **118–19**, 197
fossils 17, 41, **62–3**, 81, 81
  invertebrates 67, 207, 219, 235
  sea snakes 265
  sharks 35, 240
Fowles, John, *The French Lieutenant's Woman* 81
France 29, 110, 150, 165
  tied islands 82, 83
Franklin, John 170
Fratercula spp. **42–43**
  F. arctica (Atlantic puffin) 42
  F. cirrhata (tufted puffin) 43
  F. corniculata (horned puffin) 43
French Research Institute of the Sea (IFREMER) 13
fresh water 76, 100, 110, 170
  animals 68, 214, 220, 272
fringing reefs 206
fur 121, 175, 176
Fuss, Albert 114

G

Galápagos Islands 47, 84–5, 132, **136–7**
  birds 49, 78, 157, 189, 190
  sharks 262
  tube worms 296
Ganges Delta 110
gannets 41
garbage patches 22, 261
Garcia, Julio 274
gas bladders 118, 118, 161, 202
Gastropoda **52–5**
gastropods **52–5**, 197
gateways 25
Gecarcoidea natalis (Christmas Island red crab) 282
geogenic reefs 237
geological formations
  arches **44–7**, 171

geological formations *continued*
  cliffs 30, **40–41**, 56
  desert islands 96, **132–5**
  dunes 75, 99, **100–101**, 104
  fjords 30, **114–15**
  lagoons **76–7**, 80, 81, 100, 133, 283
  rock pools **64–5**, 237
  rocky reefs 212, **236–7**
  sea caves 45, 47, **50–51**
  seamounts 133, 248, **292–3**
  sea stacks 30, **44–7**
  tied islands 81, **82–3**
  see also beaches; trenches; volcanic activity
geysers 16
ghost sharks 289
giant oarfish **252–3**
giant tube worms **296**
Gibraltar, Rock of 83
Gibraltar, Strait of 25
Gifford, Isabella, *The Marine Botanist* 37
Ginglymostoma cirratum (nurse shark) 240–41
glaciers 115, 170–71
Glackens, William James, *Beach at Dieppe* 65
gladius 279
Glaucium flavum (yellow horned poppy) **74–5**
Glaucus atlanticus (blue sea dragon) 52, 54, 161
global conveyor belt 24–5
Globicephala melas (long-finned pilot whale) 226
Glover-Kind, John A. 86
Godiva quadricolor (nudibranch) 197
godwits 29, 127
Gondwana 16, 17
Gonostomatidae (bristlemouth) 251
gooneys 145
Gorgonian spp. (sea fan) 234
Gorgonocephalus spp. (basket star) 250
Gould, John and Elizabeth 48
grasses 122, 127
  marram 99, 101
  seagrasses 36, 37, **102–3**, 113
gravitational forces 28
Great Barrier Reef 205, 207, **210–11**, 235
Great Pacific garbage patch 22, 261
Greece, ancient 15, 209, 275, 291, 293
  mythology 41, 61, 95, 113, 221, 245, 266
  gods and goddesses 47, 70–71, 75, 75, 150, 230
  pottery 221, 237, 275
  see also Aristotle; Homer

Greece, modern 85, 236, 239
Greenland 24, 178, 180, 181
  Denmark Strait 248, 285
  fjords 115
  ice shelves and icebergs 170, 171
  mythology 182, 265
greenshanks 124
Grimpoteuthis spp. (dumbo octopus) 257, 298
groupers 210
Guam 272
guano 49
guillemots 41, 46, 115
gulfs 14, 272
Gulf Stream 158, 165
gulls 42, 127, **156–7**, 167
Günther, Albert 213
guyots 13, 133
Gygis alba (white tern) 167
Gymnothorax javanicus (giant chain moray eel) 214
gyres 22, 91, 158, 261

H

hadal zone 32–3, **298–9**
Hadley cells 142
Haeckel, Ernst 199, 218
  Kunstformen der Natur 67
Haematopus spp. 127
  H. ostralegus (Eurasian oystercatcher) 86, 127
hemoglobin 98, 190
hagfish 298
Haida people 223, 257
Hai Nei Shih Chou Chi 287
Haiti 232, 233, 266
Halichoerus grypus (gray seal) 61
Haliotis spp. (abalone) 54
hammerhead sharks 258, **262–3**
Hapalochlaena lunulata (greater blue-ringed octopus) 257
hard corals 87, 204, 205, 206
Harriotta raleighana (long-nosed chimaera) 289
hatchetfish 33
Hawai'i 30, 132, 145, 248, 290
  beaches 151, 261
  mythology 71, 86, 207, 221, 227, 243, 261, 293
  volcanoes 292
Hawaiian-Emperor seamount chain 293
hawks 137
headlands 30, 45–6, 47
Heikeopsis japonica (heikegani) 71
Hemitrygon akajei (red stingray) 245
Hensen, Victor 199
herons 70, 108, 123, 126, 127, 137

herrings 252
Heterocongrinae (garden eel) 215
Heterodontus francisci (horn shark) 243
Hexabranchus sanguineus (Spanish dancer sea slug) 135
high-tide zone 57, 84
Himantura leoparda (leopard whipray) 245
Hinds, David 290
Hinduism 55, 89, 130, 131, 151
Hippocampus spp. (seahorse) **208–9**
   H. abdominalis (big-belly seahorse) 209
   H. bargibanti (pygmy seahorse) 209
   H. guttulatus (long-snouted seahorse) 209
Hiroshige, Utagawa 50, 163
Histrio histrio (sargassum angler fish) 159
Histriophoca fasciata (ribbon seal) 61
hogfish 92–3
Hokusai 260
Holacanthus passer (king angelfish) 92–3
Holothuroidea (sea cucumber) 298
   Holothuria thomasi (tiger tail sea cucumber) 235
Homarus spp.
   H. americanus (American lobster) 69, 71
   H. gammarus (common/European lobster) 69
Homer
   The Iliad 291
   The Odyssey 162, 237, 245
Honckenya peploides (sea sandwort) 101
Humboldt Current 165
hunting industry 180, 183
   see also fishing industry
hurricanes 111, 143, **146–7**, 233
   see also tropical cyclones
Hydrocharitaceae (grasses) **102–3**
Hydrocoloeus minutus (little gull) 157
Hydrolagus melanophasma (Eastern Pacific black ghost shark) 289
Hydrophis
   H. platurus (yellow-bellied sea snake) 264
   H. schistosus (beaked sea snake) 264
   H. spiralis (yellow sea snake) 265
hydrostatic pressure 298
hydrothermal vents 248, **294–5**, 298
   life in 13, 13, 270, 296
hydrozoans 115, 161
Hydrurga leptonyx (leopard seal) **192–3**
Hypanus americanus (southern stingray) 244
Hysterocarpus traskii (tule perch) 110

I

Iatmul people 130
ibises 127
icebergs 165, **170–71**, 180, 233
   Southern Ocean 184, 186–7
ice caps 170, 180, 184
Iceland 44, 13, 233
   fjords 115
   mythology 41, 43, 47, 127
   volcanic activity 16, 85, 132
ice, sea 24, **172–3**
ice shelves **170–71**, 184
Iconographia Zoologica 98
Idiosepius notoides (pygmy squid) 278
iguanas 47, 136–7, 137
Inca Empire 49
India 86, 110, 128, 151, 283
   mythology 55, 70
Indian Ocean 12, 22, 180, **280–83**
   animals in 112, 130, 206, 250, 264
   storms 146
   tsunamis 154, 155
Indian Ocean islands 282
   Andaman Island 282, 283
   Madagascar 283
   Maldives 206, 282, 283
   Rodrigues Island 293
   Seychelles 85, 134
Indigenous peoples
   Aboriginal Australians 55, 86, 95, 101, 207, 210, 210
   of Indian Ocean islands 155, 283
   Māori 134, 214, 263
   mythology 55, 191, 214, 227, 245
   North American 91, 183, 173, 221, 223
   Canadian First Nations 53, 55
   Inuit 61, 61, 173, 176, 181, 227
   Thule 173, 182
   of Pacific Ocean islands 272
   of Russia 175
Indonesia 94, 134, 256, 259, 293
   mangroves 129
   tsunamis 155
ink 256
inlets 30, 76, 77, 115
insects 101, 123, 133, 208
International Union for the Conservation of Nature (IUCN) 275
Intertropical Convergence Zone (ITCZ) 142
Inuit peoples 61, 61, 173, 176, 181, 227
invertebrates
   insects 101, 123, 133, 208
   sponges 206, 221, 284, **290–91**
   see also cnidaria; crustaceans; echinoderms; mollusks; worms
Ireland 43, 230, 233
   Cliffs of Moher 40, 41
   mythology 151, 165
   see also Northern Ireland
Iroquois people 91
Ischadium recurvum (hooked mussel) 272
Islam 110, 227, 275, 283
islands
   barrier 76, 77, 133
   desert 96, **132–5**
   rocky reefs 212, **236–7**
   tied and tombolos 81, **82–3**
   see also Atlantic Ocean islands; Indian Ocean islands; Pacific Ocean islands
isopods 68–9, 110
Israel 239
Istiophoridae 159, **266–7**
Istiophorus albicans (Atlantic sailfish) 268–9
Italy 76, 77, 162, 237
   Capri 50, 102
   Mount Vesuvius 293
   see also Rome, ancient

J

Jacobs, Joseph 83
Janthina janthina (purple snail) 161
Japan 148–9, 279
   cuisine 67, 235
   fishing industry 259, 260, 275
   islands 132, 133, 253
   mythology 214, 226, 253, 257, 266, 266, 293
   tombolos 83
   tsunamis 155, 162
Japetella diaphana (diaphanous pelagic octopod) 251
Jardine, William, *The Naturalist's Library* 117, 117
Jaumea carnosa (marsh jaumea) 123
Java 293
Jaws (film) 243
jellyfish 76, 161, 205, **216–19**, 298
   bioluminescence 287
   as predators 54
   as prey 89, 199
jewelery 34, 55, 260
Juan Fernández islands 134
Jurassic period 17, 63, 63, 81, 81

K

Kahanamoku, Duke 214, 261
Kajikia audax (striped marlin) 269
kale, sea 73
kelp 36, 46, 57, 165, 202, 237
   forests 66, 67, **118–19**, 197
keratin 88, 225
killer whales (orcas) 35, 60, 192, 220, **222–3**
killifish 110
Kiribati 134
kittiwakes 41, 157
Kiwa hirsuta (yeti crab) 294
Kogia sima (dwarf sperm whale) 225
Köhler, Hermann Adolph, *Köhler's Medicinal Plants* 97
Kokunimasa, Utagawa 154–5
Korea 29, 55, 260
Kosteletzkya pentacarpos (saltmarsh mallow) 123
krill 69, 71, 173, 186–7, 199
Kuniyoshi, Utagawa 141, 148–9
Kuril–Kamchatka Trench 54
Kuroshio Current 150
Kwakwaka'wakw people 223

L

Labrador Current 165
Labrador Sea 14
Labridae spp. (wrasse) 237
La Fontaine, Jean de 221
lagoons **76–7**, 80, 81, 100, 133, 283
   animals in 78, 127, 206, 214
lakes 76, 233
Lamarck, Jean-Baptiste 216
lamellae 78
Lamellodysidea chlorea (Indo-Pacific blue sponge) 291
Lamna spp.
   L. ditropis (salmon shark) 240
   L. nasus (porbeagle) 159
lantern fish 251
Laridae (gull) **156–7**
   Larus dominicanus (kelp gull) 156
   Larus marinus (great black-backed gull) 157
larvae 214, 217, 237
Laticauda spp.
   L. colubrina (yellow-lipped sea krait) 267
   L. semifasciata (black-banded sea krait) 264, 264
Latimeria chalumnae (coelacanth) 35, 282
Latreutes fucorum (Sargassum shrimp) 159

INDEX

lava 19, 50, 292, 294
leafy sea dragons **212–13**
leopard seals **192–3**
Lepidochelys spp.
    L. kempii (Kemp's ridley sea turtle) 88
    L. olivacea (olive ridley turtle) 92–3
Leptonychotes weddellii (Weddell seal) 61
Leucocarbo bougainvilliorum (guanay cormorant) 49
Leucophaeus spp. (gull)
    L. atricilla (laughing gull) 156
    L. fuliginosus (lava gull) 157
Leucothea pulchra (spotted comb jelly) 284
lichen 56
lighthouses 149, 236
Limacia clavigera (orange-clubbed sea slug) 54, 55
limestone 45, 47, 50, 62, 83
Limonium vulgare (sea lavender) 123
Limosa lapponica (bar-tailed godwits) 29, 127
limpets 54–5, 65, 127, 237
Linnaeus, Carolus 289
Linuche unguiculata (thimble jellyfish) 76
lionfish 256
Liparidae (snailfish) 298
Lipophrys pholis (common blenny) 65
lithosphere 18, 300
littoral caves 50
Littorina littorea (winkle) 57
lizards *see* iguanas
Lizars, William 117
Lobodon carcinophaga (crabeater seal) 61, 173, **186–7**
lobsters 68–9, 70, 71, 300
Loligo vulgaris (European squid) 279
longshore drift 73
Lophiiformes (anglerfish) 284, 287
Lost City Hydrothermal Field 295
low-tide zone 57
Lucania parva (rainwater killifish) 110
lugworms 85, **98**
Luk, Sandy 103
Lutjanus gibbus (red snapper) 280
Lybia tessellata (pompom crabs) 70
Lythraceae 128

M

Macrocheira kaempferi (Japanese spider crab) 69
Macrocystis pyrifera (giant kelp) 118, 118, 197, 202
Macrourinae (rattail) 32, 298
Madagascar 283

maelstroms 162, 163
Magellan, Ferdinand 259–60, 259
magma 18, 294
Magnus, Olaus, *Carta Marina* 179, 265
Malcolm of Firth, John 47
Maldives 206, 282, 283
Malta 21, 46
mammals
    dolphins **220–21**, 291
    dugong **112–13**
    manatees 103, **112–13**
    narwhals 173, **178–9**, 223, 225
    orcas 35, 60, 192, 220, **222–3**
    polar bears 173, 173, **174–7**, 181
    sea lions 47, 61, **106–7**
    sea otters **120–21**
    walruses 106, 121, 173, 175, 180, **182–3**
    see also seals; whales
manatees 103, **112–13**
mangroves **128–9**, 130
Månsson, Olaf, *Carta Marina* 8–9
manta ray 33
mantis shrimp 68, 71, 282
mantle (geology) 17, 18, 132
mantle (mollusk anatomy) 54, 255, 270, 278
Māori people 134, 214, 263
    mythology 55, 191, 214, 227, 245
mapping oceans 13, 248
    early maps 8–9, 15, 110, 248, 259, 283
Marcus Aurelius 101
Mariana Trench 300
Marine Biological Laboratory (MBL) 36
Marine Megafauna Foundation 245
marlin 159, 268, 269
Marmara, Sea of 14
marram grass 99, 101
marshes 78, 108, 111, 127
    salt marshes **122–3**
matrilineal societies 223
Mauritania 23, 239
Mayan culture 55, 89, 143
McTaggart, William, *Amongst the Bents* 99
meadows, seagrass 102, 103, 113
medicines 75, 96, 98, 202, 209, 287
Medieval bestiaries 226
Mediterranean Sea 15, 25, 28, 83, 239
    animals and plants in 55, 103, 197, 266
Medusozoa spp. (jellyfish) **216–19**
Megadyptes antipodes (yellow-eyed penguin) 190, 191
Megalodon 34

Megaptera novaeangliae (humpback whale) 159, 226, **228–9**
Megatrygon microps (smalleye stingray) 245
Melanocetus johnsonii (humpback anglerfish) 285
melons 117
Melville, Herman, *Moby Dick* 226–7, 269
menopause 223
mermaids 113, 261
Mesonychoteuthis hamiltoni (colossal squid) 278
mesopelagic (twilight) zone 250
methane seeps 248
Mexico 55, 85, 121, 246–7
Mexico, Gulf of 14, 29, 31, 55
Microcarbo niger (little cormorant) 49
microorganisms 199
    archaea 17, 298
    phytoplankton 173, 181, 197, 198, 199
    see also algae; bacteria
microplastics 34
Mid-Atlantic Ridge 13, 232, 295
middle-tide zone 57
mid-ocean ridges 12, 18, 294
Mid-Atlantic Ridge 13, 232, 295
migrations
    bird 77, 124–5, 127, 134, 167, 173
    crab 282, 282
    diel vertical migration (DVM) 33, 250
    fish 115, 214, 268, 275
    human 96, 143
    turtle 89
    walrus 182
    whale 159, 226
Millot, Adolphe Philippe 203
Milton, John, *Paradise Lost* 49
mimicry 256
minerals 63, 84–5, 294
mining 181, 249
Minoan culture 257
Mirounga leonina (Southern elephant seal) 61, 184, 192
mists **164–5**
Mitsukurina owstoni (goblin shark) 242
Mobula spp. 33
    M. munkiana (Munk's devil rays) 246–7
Moche people 91
Moken people 155, 283
Mola mola (ocean sunfish) 219
Mollusca (mollusks) 53
mollusks 17, 56, 62, 63
    bivalves 53, 127, **270–73**
    gastropods **52–5**, 197
    see also cephalopods

mollymawks 145
Monacanthidae (filefish) 158
Monachus monachus (monk seal) 51
Mondrian, Piet 101
Monet, Claude, *Impression, Sunrise* 165
monkeys 108
Monodon monoceros (narwhal) 173, **178–9**
monsoons 27, 280
Monteith, James 143
moon 28
moray eels 102, 214
mother-of-pearl 271, 279
mounds 22, 285
mountains (seamounts) 133, 248, **292–3**
mudflats 108, **122–3**
mudskippers 108
Munch, Edvard, *The Scream* 293
Murayama, Hashime 269
Murex spp. (sea snail) 54
murres 46
mussels 57, 270, 272
Myanmar 283
Myctophidae (lantern fish) 251
Myliobatoidei spp. (stingray) **244–5**
Myopsida (squid) **278–9**
Mysticeti (baleen whale) 225
mythology
    African 233, 233, 266
    Australian 55, 86, 95, 207, 210
    Asian 55, 130, 155, 253, 269
    Chinese 91, 261
    Indian 55, 70
    Japanese 214, 226, 253, 257, 266, 266, 293
    Pacific islanders 241, 257, 260, 261, 272, 275
    creatures 123, 301
    mermaids 113
    sea monsters 8–9, 255, 279
    sea serpents 253, 264, 265–6, 266
    unicorns 178, 178, 179
    European 67, 239
    British 81, 146, 165
    Icelandic 41, 43, 47, 127
    Irish 41, 151, 165
    Norse mythology 151, 165, 165, 226, 266, 279
    Roman 209, 293
    see also Greece, ancient
    Māori 55, 191, 214, 227, 245
    North American 91, 173, 221, 223
    Canadian First Nations 53, 55
    Greenland 182, 265
    Hawaiian 71, 86, 207, 221, 227, 243, 261, 293
    Inuit 61, 61, 176, 227

mythology *continued*
    see also religions
Mytilida (mussel) 57
Myxinidae (hagfish) 298

N

nacre 53, 55, 271, 272
Namibia 23, **104–5**, 165
Nannopterum spp.
    N. auritus (double-crested cormorant) 49, 49
    N. harrisi (flightless cormorant) 49
Napoleon I 233
narwhals 173, **178–9**, 223, 225
Nasalis larvatus (proboscis monkey) 108
natal homing 89
Natator depressus (flatback sea turtle) 90
Naticidae (moon snails) 54
nautiluses 257
Nazaré Canyon **152–3**
Nazca culture 94, 222
Neap tides 28
Negaprion brevirostris (lemon shark) 128
nematocysts 54, 217–19
Nenet people 181
Neolithodes diomedeae (king crab) 297
Neomonachus schauinslandi (Hawaiian monk seal) 61
Nephropidae spp. (lobsters) 70
nerve nets 216, 235
Netherlands, the 101, 142
New Guinea 55, 94, 143
Newton, Isaac 28
New Zealand 115, 134, 162
    animals in 145, 190, 191
    coastline 46, 50, 56–7, 73
    Māori people 134, 214, 263
        mythology 55, 191, 214, 227, 245
Nigeria 54, 233
Nile Delta 110
Nodder, Frederick 159
Norse mythology 151, 165, 165, 226, 266, 279
North Atlantic Current 158
North Atlantic Equatorial Current 158
Northern Ireland 51
Norway 162, 162, 181, 265
    animals in 43, 46, 49, 176
    fjords **114–15**
Norwegian Sea 15
notothenioids 181
nudibranchs 197

O

oarfish, giant **252–3**
ocean basins **12–13**
    see Arctic Ocean; Atlantic Ocean; Indian Ocean; Pacific Ocean; Southern Ocean (Antarctic Ocean)
ocean currents 140, 143, 162, 165
    in Atlantic Ocean 104, 158, 230
    deep-water **24–5**, 284, 298
    formation of garbage patches 260
    in Pacific Ocean 137, 150
    in Southern Ocean 150, 184
    surface **22–3**, 26, 81, 82, 108
ocean evolution **16–19**, 37
ocean floor 13, 18, 25, 205, **248–9**
    in the dark zone 284–5
    plants 103, 118
    see also hydrothermal vents
ocean gateways 25
Oceania 259
oceanic crust 12, 18, 294
ocean trenches 12, 248, 298, **300–301**
    Atlantic Ocean 230
    Pacific Ocean 54, 300
ocean zones 13, **32–3**
    abyssal 268, **298–9**
    dark **284–5**
    hadal **298–9**
    sunlit **196–7**
    twilight **250–51**, 284, 285
    see also ocean floor; ocean trenches
Octopoda **254–7**
    Octopoteuthis deletron (octopus squid) 279
    Octopus pallidus (pale octopus) 256
    Octopus vulgaris (common octopus) 256
    Octopus wolfi (pygmy octopus) 255
octopods 251
octopuses **254–7**, 278, 298
Ocypode gaudichaudii (painted ghost crabs) 85
Odobenus rosmarus (walrus) 173, **182–3**
Odontoceti (toothed whales) 225
Odontodactylus scyllarus (peacock mantis shrimp) 282
Oegopsida (squid) **278–9**
offshore wind farms 142, 143
Onychoprion fuscatus (sooty tern) 167
operculum 54, 289
Ophiothrix fragilis (common brittle star) 237
orcas 35, 60, 192, 220, **222–3**

Orcinus orca (orca) 220, **222–3**
Ortelius, Abraham 259
Osedax (zombie worm) 296
osteichthyes (bony fish) 252
Ostreida (oysters) 108
Otariidae spp. (sea lions) **106–7**
    Otaria byronia (South American sea lion) 107
otters, sea **120–21**
Outerson, William 35
outgassing 16
overfishing 91, 245, 275, 290
Oxudercinae (mudskippers) 108
oystercatchers 86, 123, 127
oysters 63, 65, 108, 270, 271, 272

P

Pacific Ocean 12, 30, 181, **258–61**
    animals in 42, 67, 113, 121, 204, 264, 280
    storms 146, 146
Pacific Ocean islands 258, 260, 260
    Bora Bora 77, 196
    Easter Island 167
    Fiji 204, 241, 257, 261
    Palau 132, 272
    Polynesian people 134, 259, 261, 261, 275, 293
    Samoa 96
    Solomon Islands 258, 260, 261, 290
    see also Galápagos Islands; Hawai'i
Pacific Ring of Fire 18, 293, 300
Pagophila eburnea (ivory gull) 156
Paguroidea (hermit crab) 68, 70
Pagurus bernhardus (common hermit crab) 65
Palaeophis colossaeus 265
Palau 132, 272
palm trees **96–7**, 133
pancake ice 173
Pangaea 16
Panthalassic Ocean 16
Papua New Guinea 130, 234
Paracentrotus lividus (purple sea urchin) 67
Paralabrax clathratus (kelp bass) 118–19
parrotfish 85, 87, 206
Patellogastropoda (limpet) 54
    Patella vulgata (common limpet) 55, 65
Paul the Deacon 162
Payne, Roger 34, 226
pearls 260, 260–61, 270, 272
pebbles 73, 81, 84
Pectinidae (scallop) 272
pedicellariae 67

pelagic octopods 251
Pelecanus spp. (pelicans)
    P. conspicillatus (Australian pelican) **94–5**
    P. erythrorhynchos (American white pelican) 123
    P. occidentalis (brown pelican) 94
    P. thagus (Peruvian pelican) 94
pelicans **94–5**, 123, 123
penguins 115, 137, 185, **188–91**, 192
pen shells 271
perches 110
periwinkles 55
permineralization 63
Persian Gulf 272
Peru 23, 49, 91, 94
petrels 158
Phaeophyceae spp. (brown kelp) 165
Phalacrocoracidae (cormorants) **48–9**
Phalacrocorax spp. (cormorants) 48, 48
    P. carbo (great cormorant) 48
    P. punctatus (spotted shag) 56–7
Philippines 76, 96, 261, 275, 300
Philippine Sea 15
Phocarctos hookeri (New Zealand sea lion) 107
Phocidae (seals) **60–61**
Phoebastria nigripes (black-footed albatross) 145
Phoenicians 34, 209
Phoenicopteridae spp. **78–9**
    P. andinus (Andean flamingo) 79
    P. roseus (greater flamingo) 79
    P. ruber (American or Caribbean flamingo) 79, 79
photophores 251, 287
photosynthesis 20, 197, 202, 206, 207, 250
Phycodurus eques (leafy sea dragon) **212–13**
Phyllorhiza punctata (Australian spotted jellyfish) 217, 219
Phyllospadix (surf grass) 103
Physalia physalis (Portuguese man-of-war) 54, 161
Physeter macrocephalus (sperm whale) 33, 224, 228–9, 279
phytoplankton 23, 173, 181, 197, 198, 199
Pierre, André 232
Pinna nobilis (noble pen shell) 271
Pinnidae (pen shell) 271
pirates 51, 133, 137, 233
Pisaster giganteus (giant-spined star) 119

INDEX

Planes minutus (Sargassum crab) 159
plankton 12, 23, 23, 37, **198–9**
    phytoplankton 173, 181, 197, 198, 199
    zooplankton 68, 181, 197, 199, 250
plants **36–7**
    coconut palms **96–7**, 133
    grasses 122, 127
        marram 99, 101
        seagrasses 36, 37, **102–3**, 113
    mangroves **128–9**, 130
    poppies **74–5**
    salt marshes 122, 122, 123
    tides effect on 29
    see also seaweeds
plastics 22, 34, 37, 261
Platax teira (longfin batfish) 280
Platichthys flesus (flounder) 249
Plato 239
Plectorhinchus polytaenia (ribboned sweetlips) 280
Pliny the Elder 90, 216, 217, 221, 241, 287
Pliosaurus spp. 41
plovers 73, 85, 101, 127
Pluvialis spp.
    P. apricaria (European golden plover) 127
    P. squatarola (black-bellied plover) 127
pods 116, 117, 178–9, 221, 223
poisonous animals 71, 272, 297
    see also venomous animals
polar bears 173, 173, **174–7**, 181
polar cells 142
polar regions 170–71, 172–3
    Antarctic *see also* Antarctica; Southern Ocean (Antarctic Ocean)
    Arctic 170, 172, 175, 176, 182
    see also Arctic Ocean
pollution 86, 207, 211, 261
Polo, Marco 101
Polynesian people 134, 259, 261, 261, 275, 293
polyps 161, 205, 206, 210, 217, 235
poppies **74–5**
porbeagles 159
Porifera (sponges) **290–91**
Poromya granulata (clam) 272
Porpita porpita (blue button jellyfish) 161
porpoises 220
Portugal 72, 152–3
Posidoniaceae (grasses) **102–3**
    Posidonia oceanica (Neptune grass) 103
Pottle, Derek 173
power stations 29

prehistoric life 17, 34–5, **62–3**
Prêtre, Jean-Gabriel 160
Prionurus punctatus (yellowtail surgeonfish) 92–3
Procellariiformes 144
Pseudalsophis biserialis (Galápagos racer snake) 137
Pseudocolochirus violaceus (sea apple) 235
Pseudoliparis spp.
    P. amblystomopsis (hadal snailfish) 300
    P. swirei (Mariana snailfish) 298
Pseudomogoplistes vicentae (scaly cricket) 73
Pterois (lionfish) 256
Ptolemy, *Geography* 15
Puerto Rico Trench 230
puffins 41, **42–3**
Pusa spp.
    P. hispida (ringed seal) 61
    P. sibirica (Baikal seal) 61
Pycnopodia helianthoides (sunflower sea star) 235
Pygoscelis spp.
    P. adeliae (Adélie penguin) 189
    P. antarctica (chinstrap penguin) 189, 191
    P. papua (gentoo penguin) 189, 191
Pyrocystis lunula (algae) 32

R

radial symmetry 66, 235
radulas 54, 279
rainstorms 280
Raphus cucullatus (dodo) 282
rattails 32, 298
rays
    manta ray 33
    stingrays 196, **244–5**, 246–7
Read, Mary 133
Recurvirostra spp. (avocet) 127
Red Sea 21, 51, 282
redshanks 127
reefs 212, **236–7**
    coral 133, 197, **204–7**, **210–11**, 235, 282
Regalecus glesne (giant oarfish) **252–3**, 265
religions 128, 143, 146
    Christianity 227, 272
    Hinduism 55, 89, 130, 131, 151
    Islam 110, 227, 275, 283
    Shinto 147, 147
renewable energy 29, 142, 143
reptiles
    crocodiles 108, **130–31**
    Galápagos Islands 137

sea snakes 256, **264–7**
sea turtles 85, **88–91**, 92–3, 133, 219
Rhincodon typus (whale shark) 240, 242, 282
Rhinomuraena quaesita (ribbon eel) 214
Rhizophoraceae (mangrove) 128
Rhizostomeae (jellyfish) 219
Riftia pachyptila (giant tube worm) 294, **296**
Ring of Fire, Pacific Ocean 18, 293, 300
Rissa tridactyla (kittiwakes) 157
rivers 30, 31, **108–11**, 130
rivulets 109
rock pools **64–5**, 237
rocky beaches **56–7**, **72–3**, 127
rocky reefs 212, **236–7**
Rodrigues Island 293
rogue waves 150–51, 260
Rome, ancient 15, 51, 275, 238, 239, 293
    medicines 202, 209, 291
    mythology 209, 293
Roosevelt, Theodore, *Book-Lover's Holidays in the Open* 95
root systems 128
Rumi, *Divani Shamsi Tabrizi* 271
Ruskin, John 189
Russell, Isaac 75
Russia 116, 117, 175, 178, 180, 181
Rudolf Island 45

S

sailfish 268, 268–9
Saint Helena 233
Saint Pierre and Miquelon 83
Sakamoto, Ryuichi 155
salemas 137
Salicornia europaea (marsh samphire) 122
salinity 21
    see also saltwater
salp 199
Salpidae (salp) 199
Salter, James 110
salt marshes 108, **122–3**
saltwater 12, **20–21**, 24, 76, 89, 110
saltwater crocodiles **130–31**
saltwater wetlands 124–5
Sami people 181
Samoa 96
samphire 122, 122
sand 84–5, 84
sand dollars 67
sand dunes 75, 99, **100–101**, 104
sandhoppers 69, 126
sandpipers 124, 126, 127

sandstone 40, 62
sandy beaches **84–7**
San people 104
Sargasso Sea 14, 36, **158–9**, 214
Sargassum (seaweed) 158, 197
Sargent, John Singer, *Atlantic Storm* 231
Sarotherodon melanotheron (blackchin tilapia) 76
satellite imaging 13, 111, 198, 248, 298
Saudi Arabia 283
Scales, Helen 299
scallops 270, 271, 272
Scandinavia 8–9, 180, 279
    see also Denmark; Iceland; Norway
Scarinae (parrotfish) 206
Schmidt Ocean Institute 35, 249
Schussele, Christian 200–201
Scleractinia (corals) 207
Scolopacidae (waders) **126–7**
Scotia Sea 14
Scotland 43, 61, 165, 237
    coastline 51, 83, 237
    fjords 115
Scotoplanes (sea pig) **297**, 298
scrimshaw 225
scutes 88
sea anemones 57, 70, 161, 205, 216, 237
sea apples 235
seabed *see also* ocean floor
sea caves 45, 47, **50–51**
sea cows (manatees) 103, **112–13**
sea cucumbers 235, 235, 284, 286, **297**, 298
sea dragons 208, **212–13**
sea fans 234
seagrasses 36, 37, **102–3**, 113
seagulls 42, 127, **156–7**, 167
seahorses **208–9**, 212–13
sea ice 24, **172–3**
sea kale 73
sea kraits 264, 267
sea lettuce 202
sea levels 30, 81, 82, 86, 207, 210
    see also tides
sea lilies 235, 284
sea lions 47, 61, **106–7**
seals 51, **60–61**, 85, 105, 181
    crabeater seals 61, 186–7
    elephant seals 61, 184, 192
    leopard seals **192–3**
    as prey 105, 175–6, 222, 240
    and sea lions 106–7
sea mists **164–5**
sea monsters 8–9, 81, 255, 279
    serpents 253, 264, 265–6, 266
seamounts 133, 248, **292–3**

sea nettles 216–17, 219
sea otters **120–21**
sea pigs **297**, 298, 300
sea poppies 75
seas **14–15**
    Sargasso Sea 14, 36, **158–9**, 214
    see also Mediterranean Sea
sea sandwort 101
sea serpents 253, 265–6, 265, 266
sea slugs 52, 54, 55, 135, 161
sea snails 54, 55, 56, 123, 161, 237
sea snakes 256, **264–7**
sea sponges 206, 221, 284, **290–91**
sea spurge 101, *101*
sea squirts 291
sea stacks 30, **44–7**
sea stars (starfish) 67, *119*, 235, 237
sea stumps 46
sea turtles 85, **88–91**, 92–3, 133, 219
sea urchins 35, 57, 65, **66–7**, 235, 237
    as prey *119*, 121
seawater **20–21**
    see also saltwater
seaweeds 36, 56, **202–3**, 237
    Sargassum (seaweed) 158, 197
    see also kelp
sediment 25, 31, 249
    pebbles 73, 81, 84
    sand 84–5, *84*
sedimentary rock 41, 62
seed dispersal 128
semidiurnal tides 29
Sentinelese people 283
Sepioteuthis spp.
    S. lessoniana (bigfin reef squid) 278, *279*
    S. sepioidea (Caribbean reef squid) 278, *278*
septa 207
Seri people 91
sessile crustaceans 69
Seychelles 85, 134
Shackleton, Ernest 184, *185*
shags 48, 56–7
Shakespeare, William 49, 143
    Tempest 58–9
shale 62
sharks 47, 108, 181, **240–43**, 260, 289
    great white sharks 159, 222–3, 243
    hammerhead sharks 258, **262–3**
    Megalodon 34
    nurseries 76, 128
    tiger sharks 276–7
    whale sharks 240, 242, 282
shellfish *see* crustaceans; echinoderms; mollusks
shells 34, 53–4, 85

Shinto 147
ships 110, 143
    pollution from 86
    shipwrecks 81, 171, 172, 249, 282
    in art 58–9, 302–3
    Skeleton Coast 105, *105*
    see also explorers
shoals 275
shores, rocky **56–7**
shrimp 68, 71, 78, 123, 159, 282
Singapore 269
Sinistrofulgur perversum (lightning whelk) 55
siphonophores 265
siphons 255
sirenians (sea cows) 112–13
Sirenia (sea cows) **112–13**
skates 243, 289
Skeleton Coast, Namibia **104–5**, 165
skimmers 157
skuas 42, 157
slave trade 230, 233
slugs 53
    see also sea slugs
Smith, Joseph 53
smugglers 51, 81, 133
snailfish 298, 300
snails 53
    see also sea snails
snakes 137, 264
    see also sea snakes
snappers 280
Snorri Sturluson 115
soft corals 204, 205, 206
Solomon Islands 258, 260, 261, 290
Sommerville, James M, *Ocean Life* 200–201
Somniosus microcephalus (Greenland shark) 181
sonar (sound navigation and ranging) 13, 248
South Africa 34, 150
Southern Ocean (Antarctic Ocean) 12, 24, **184–5**, 186–7
    ice 170–71, *173*
    storms 150
spade fish 276–7
Spain 83, 86, 103, 109, 150
sparrows 123
Spartina (cord grass) 122
spearfish 268, 269
Sphenisciformes (penguins) **188–91**
    Spheniscus mendiculus (Galápagos penguin) 137, *189*
Sphyraena barracuda (great barracuda) 33
Sphyrnidae **262–3**
    S. lewini (scalloped hammerhead) 263

Sphyrnidae *continued*
    S. mokarran (great hammerhead) 263
    S. tiburo (bonnethead) 263
    S. zygaena (smooth hammerhead) 263
spicules 290
Spielberg, Steven 243
spits 81
Spondylus (thorny oyster) 272
sponges 206, 221, 284, **290–91**
spookfish 284, 289
spoonbills 127
sports 151, 237
spray zone 57
Spring tides 28
squid 32, 34, 251, **278–9**, 287
starfish 67, *119*, 235, 237
Stegostoma tigrinum (zebra shark) 242
Stenella longirostris (spinner dolphin) 220, *221*
Sterna paradisaea (Arctic tern) 167
Sternidae (terns) 85, **166–7**
Sternoptychidae (hatchetfish) 33
stick insects 135
stingrays 196, **244–5**, 246–7
stipes 118
stomata 99
Stomatopoda (mantis shrimp) 68, 71
storks 108
storms 27, 81, 150
    see also tropical cyclones
storm-tide zone 84
stromatolites 37
Strongylocentrotus purpuratus (Pacific purple sea urchin) 67
Stygiomedusa gigantea (giant phantom jelly) 298
Stygiotantulus stocki 69
subduction zones 19
submergent coastlines 30
submersibles 13, 249, 298, 300
    ALVIN 34, 269
    Deepsea Challenger 299
Sun 26, 28
Sundarbans 128
sundials 54
sunfish 219
sunken cities **238–9**
sunlit zone 32–3, **196–7**
supervolcanoes 293
surface currents **22–3**, 24, 26
surface features
    icebergs 165, **170–71**, 180, 233
    ice shelves **170–71**, 184
    mists **164–5**
    sea ice 24, **172–3**
    tropical cyclones 143, **146–7**, 260, 280

surface features *continued*
    tsunamis **154–5**, 162, 253, 260, 293, 300
    whirlpools **162–3**
    see also waves; wind
surfing 110, 151, **152–3**, 261
surgeonfish 92–3
Surtsey 132
sustainable materials 37, 103
Suzuki, Shunryu 150
swallower fish 32, 284
swamps 76, 126, 130
sweetlips fish 280
swells 149, 150
swim bladders 268, 298
swordfish 268, 269
symbiosis 92–3, 206, 272, 296

T

Taeniura lymma (bluespotted ribbontail ray) 245
Tasmanian Passage 25
Tasman Sea 15
tawaki 191
Teach, Edward (Blackbeard) 87, 233
tectonic plates 17, **18–19**, 30, 293
    formations of islands 132, 134, 136
    and fossil exposure 17
    underwater features, formation of 249, 294, 300
temperature
    air 140
    and climate 26
    sea 21, 146, 173, 199, 211, 280, 296
    and sex, in turtles 91
Tenniel, John 183
terns 85, 101, 108, 127, 157, **166–7**
terrapins 123
Tethys Ocean 17
Tetrapturus angustirostris (shortbill spearfish) 269
Thailand 47, 155, 283
Thalasseus bernsteini (Chinese crested tern) 167
Thaumoctopus mimicus (mimic octopus) 256
Théel, Johan Hjalmar 297
thermals 145
thermohaline circulation (THC) 24–25
Threskiornithidae (wader) 127
Thule people 173, 182
Thunnus spp. (tuna) **274–5**
    T. albacares (yellowfin tuna) 274

Thunnus spp. (tuna) *continued*
   T. maccoyii (southern bluefin tuna) 274
   T. obesus (big eye tuna) 274
   T. thynnus (Atlantic bluefin tuna) 275
tidal pools (rock pools) **64–5**, 237
tides **28–9**, 148
   energy generated from 29
   mangroves 128
   whirlpools, formation of 162
tied islands 81, **82–3**
Timor 130
Titanic 171
Tolstoy, Fyodor 75
tombolos 81, **82–3**
Tonna galea (giant tun) 55
toothed whales 178, 223, 225, 226
tornadoes 162
tortoises 89, 134, 136, 137, 282
totipalmate feet 48, 94
tourism 85–6, 87, 114, 162, 188, 274
trade winds 142, 230
transportation (geology) 25, 31
tree of life 96
trenches 12, 248, 298, **300–301**
   Atlantic Ocean 230
   Pacific Ocean 54, 300
Trichechus spp. (manatees)
   T. inunguis (Amazonian manatee) 113
   T. manatus (West Indian manatee) 112, 113
   T. senegalensis (African manatee) 113
Tridacna gigas (giant clam) 270
trilobites 35, 248
Tringa spp.
   T. guttifer (spotted greenshank) 124
   T. totanus (common redshank) 127
Tripolium pannonicum (sea aster) 123
Tristan da Cunha 233
Trivett, John 175
tropical cyclones 143, **146–7**, 260, 280
tsunamis **154–5**, 162, 253, 260, 293, 300
tube feet 66, 235, 297
tube worms 294, **296**
tuna **274–5**
tun shells 55
Türkiye 230
Turner, J. M. W.
   Vesuvius in Eruption 293
   The Wreck of a Transport Ship 302–3
turnstones 73, 126
Turritopsis dohrnii (immortal jellyfish) 217
Tursiops truncatus (bottlenose dolphin) 221

turtles, sea 85, **88–91**, 92–3, 133, 219
tusks 113, 175, 178, 179, 182
twilight zone 32–33, **250–51**, 284, 285
typhoons *see* tropical cyclones
Typhoon Tip 27
Tyrian purple 34, 55

U

Ulva lactuca (sea lettuce) 202
underwater features
   barrier reefs 205, 206, 207, **210–11**, 235
   currents **24–5**, 284, 298
   hydrothermal vents 248, **294–5**, 298
   seamounts 133, 248, **292–3**
   waterfalls 248, 285
unicorns 178, *178*, 179
United Nations (UN) 134, 207
United States of America 164; 181
   Alaska 43, 111, 180
   Aquarius Reef Base 239
   beaches 84, 86, 87
   coastline 46, 50, 123, 237
   fossils 62
   Indigenous peoples 91, 183, 173, 221, 223
   islands 51, 133
   Mississippi River 31, 110, 111
   navy 117, 226
   whirlpools 162
   see also Hawai'i
upwelling 23, 104–5, 137
urchins, sea 35, 57, 65, **66–7**, 235, 237
   as prey 119, 121
Uria aalge (guillemot) 46
Urile pelagicus (pelagic cormorant) 46
Ursus maritimus (polar bear) 173, **174–7**

V

Vampyroteuthis infernalis (vampire squid) 287
van Leeuwenhoek, Antonie 35, 199
Velella velella (by-the-wind sailor) **160–61**
venomous animals
   invertebrates 54, 67, 217–19, 255
   fish 245, 289
   sea snakes 264, *264*, 266

vents, hydrothermal 248, **294–5**, 298
   life in 13, *13*, 270, 296
Verne, Jules, *Twenty Thousand Leagues Under the Sea* 19, 37, 239, 279
viperfish 33, 284, 285, 287
volcanic activity
   hydrothermal vents 248, **294–5**, 298
   mid-ocean ridges 12, 18, 232, 294, 295
   sea caves, formation of 50
   seamounts 133, 248, **292–3**
   volcanoes 17, 132, 134, 239, 248, **292–3**, 300
vortexes 162

W

waders (birds) 110, **126–7**
Wales 157
Walker, Boyd 253
walruses 106, 121, 173, 175, 180, **182–3**
warm currents 22
Watasenia scintillans (firefly squid) 32, 251, 279
water **20–21**
   see also fresh water; saltwater
waterfalls 233
   undersea 248, 285
Waterhouse, John William, *Miranda* 58–9
water pressure 33, 284, 297
waterspouts 143
wave-cut platforms 41
waves **148–51**, 152–3
   and rock formations 41, 45–6, 56, 73, 73
   sandy beaches, formation of 84
   seaweed effect on 118
   see also tsunamis
weather
   climate **26–7**
   mists and fogs **164–5**
   monsoons 27, 280
   tropical cyclones 143, **146–7**, 260, 280
   tsunamis **154–5**, 162, 253, 260, 293, 300
   see also waves; wind
weathering *see* erosion
wetlands 76, **122–3**, 124–5, 197
whales 34–5, **224–7**, 296, 297
   beluga **116–17**, 173, 178, 225
   blue 12, 226, 258
   bowhead 173
   diet 69, 105, 137, 185
   humpback 159, 228–9, 258
   pilot 226

whales *continued*
   sperm 33, 228, 279
   see also dolphins; orcas
whelks 55, *55*
whirlpools **162–3**
whiskers 61
white smokers 294
Widder, Edith, *Below the Edge of Darkness* 250
wind 22, **140–43**, 146, 176
   and waves 148–9
   see also erosion
wind farms 142, 143
winkles 57
Wood, John George, *The Illustrated Natural History Vol.2* 176
woodlice 68
worms 103, 110, 123, 294
   lugworms 85, **98**
   tube worms 294, **296**
wrasse 237, 261

X

Xenocys jessiae (black-striped salemas) 137
xenophyophores 300
Xestospongia muta (giant barrel sponge) 291
Xiphiidae (billfish) **268–9**

Y

Yidinji people 207
Yoshitoshi Taiso 147
Yukon Delta 111

Z

Zalophus wollebaeki (Galápagos sea lion) 47
Zambia 233
Zheng 283
zodiac 69, 70, 71
zones 13, **32–3**
   abyssal 268, **298–9**
   dark **284–5**
   hadal **298–9**
   sunlit **196–7**
   twilight **250–51**, 284, 285
zooplankton 23, 68, 181, 197, 199, 250
zooxanthellae 161, 206, 207, 211, 270
Zosteraceae (grasses) **102–3**
   Z. marina (common eelgrass) 103

Acknowledgments

Dynamo would like to thank the following people for their help with making this book:

Editing: Nick Funnell, Belinda Gallagher

Editorial support: Sara Stanford, Danni Turner

Designing: Annie Arnold, Kate Ford, Simon Oliver, Matthew Taylor

Proofreading: Clara Penn

Indexing: Elizabeth Wise

Additional picture research: Sarah Hopper

DK would like to thank the following people for their help with making this book:

Fact-checking: Michelle Harris, Katherine Kling, Priyanka Lamichhane

Editorial assistance: Peter Francis

In Delhi:

Senior Jacket Designer: Suhita Dharamjit

Art Editor: Arshti Narang.

The publisher would like to thank the following for their kind permission to reproduce their photographs:

(Key: a-above; b-below/bottom; c-centre; f-far; l-left; r-right; t-top)

1 Getty Images: Sepia Times. **2-3 Dreamstime.com:** Aleksandr Veselkov (circle of shells). **2 Alamy Stock Photo:** vivacity.tv (cr). **Dorling Kindersley:** Colin Keates / Natural History Museum, London (tr). **3 Alamy Stock Photo:** vivacity.tv (cr). **Dorling Kindersley:** Ruth Jenkinson / Holts Gems (cra); Stephen Oliver (bl). **4-5 Alamy Stock Photo:** ADDICTIVE STOCK CREATIVES. **8-9 Wikipedia:** Niday Picture Library. **10-11 Warren Keelan Photography**. **13 Getty Images:** BERTRAND LANGLOIS / AFP (clb). **naturepl.com:** Alex Mustard (tr). **Science Photo Library:** PACIFIC RING OF FIRE 2004 EXPEDITION. NOAA OFFICE OF OCEAN EXPLORATION; DR. BOB EMBLEY, NOAA PMEL, CHIEF SCIENTIST (br). **14 Alamy Stock Photo:** Oliver Hoffmann (bc). **Shutterstock.com:** Massimiliano Finzi (bl); Tarpan (c); Pugalenthi Iniabarathi (cr); TLF Images (br). **15 Alamy Stock Photo:** Oleksandr Rupeta (bl); Science History Images (tc). **Dreamstime.com:** Anastas Dimitrov (cr). **Getty Images / iStock:** Nicola Micheletti (cl); Mumemories (br). **Shutterstock.com:** PhotoRoman (bc); Roman Sigaev (c). **16 Getty Images / iStock:** kotangens (br). **17 Shutterstock.com:** MarcelClemens (tr). **19 BluePlanetArchive.com:** Doug Perrine (b). **21 Alamy Stock Photo:** imageBROKER.com GmbH & Co. KG (tr); Adrien Ledeul (bl). **22 Shutterstock.com:** Rich Carey (cra). **23 naturepl.com:** Ralph Pace / Minden (cra). **24 Dreamstime.com:** Martinmark (bl). **25 MBARI** (cra); **Science Photo Library:** JAXA / NASA / SCIENTIFIC VISUALIZATION STUDIO (cla); PLANETARY VISIONS LTD (br). **27 Alamy Stock Photo:** Photo 12 (br). **Getty Images / iStock:** petesphotography (tr). **29 Alamy Stock Photo:** Xinhua (cr). **naturepl.com:** David Tipling / 2020VISION (cl). **30 BluePlanetArchive.com:** Masa Ushioda. **31 Getty Images:** Abstract Aerial Art (cr). **Science Photo Library:** NASA (bl). **32 naturepl.com:** Brian Skerry / Minden (c); Norbert Wu / Minden (crb); Solvin Zankl (bc). **Science Photo Library:** Frank Fox (cra). **33 Alamy Stock Photo:** blickwinkel (tr); mauritius images GmbH (cb). **Dreamstime.com:** Aquanaut4 (cla); Richard Brooks (cra); Diego Grandi (cr). **naturepl.com:** David Shale (cl). **34-35 Science Photo Library:** Roman Uchytel (tc). **34 Alamy Stock Photo:** Kate After (cb); Heritage Image Partnership Ltd (cl). **Bridgeman Images:** The Stapleton Collection (cr). **Getty Images:** by wildestanimal (bc). **Science Photo Library:** A. James Gustafson (cla). **35 Alamy Stock Photo:** Science History Images (cl); The Natural History Museum (cr); Universal Images Group North America LLC (clb); The Natural History Museum (cb). **Dreamstime.com:** Pablo Caridad (cra); Musat Christian (bl). **Schmidt Ocean Institute** (br). **Science Photo Library:** Marek Mis (tc). **36 Alamy Stock Photo:** Associated Press (clb); Science History Images (cra); The Natural History Museum (crb); Seaphotoart (bl). **naturepl.com:** David Fleetham (br); Alex Mustard (cla). **Science Photo Library:** Steve Gschmeissner (cl). **37 Alamy Stock Photo:** Scenics & Science (clb); Maria Tolkacheva (ca); The History Collection (cl); SuperStock (cr). **naturepl.com:** John Cancalosi (tr). **NOAA:** Daniel Wagner (bl). **Plastic Soup Foundation** (br). **Science Photo Library:** Richard Bizley (tc). **Wikipedia:** Isabella Gifford (cb). **38-39 Alamy Stock Photo:** JPTenor. **40 Getty Images:** Peter Unger. **40-41 Shutterstock.com:** ArtMari (background). **41 Getty Images:** Art Media / Print Collector (tl). **naturepl.com:** Vladimir Medvedev (bl). **42 Getty Images / iStock:** Alphotographic (ca). **42-43 Dreamstime.com:** Dzmitry Auramchik (background). **43 Alamy Stock Photo:** Vintage Travel and Advertising Archive (tc). **Kevin Morgans:** (bc). **naturepl.com:** Jacob S. Spendelow / BIA / Minden (cr); Gerrit Vyn (crb). **44-45 Getty Images:** James Kelly-Smith / 500px. **45 Alamy Stock Photo:** Quagga Media (tc). **Getty Images / iStock:** Yakovliev (background). **46-47 Getty Images / iStock:** Yakovliev (background). **46 Alamy Stock Photo:** robertharding (ca). **47 Jakub Fryš** (tr). **48-49 Getty Images / iStock:** Thomas Faull (background). **48 Alamy Stock Photo:** Old Images. **49 Alamy Stock Photo:** Rick & Nora Bowers (cr). **Bridgeman Images:** Freer Gallery of Art, Smithsonian Institution / Gift of Arthur M. Sackler (bc). **naturepl.com:** Henley Spiers (tc); Doug Wechsler (cra). **50-51 Shutterstock.com:** Sabelskaya (background). **50 Bridgeman Images:** San Diego Museum of Art / Bequest of Mrs. Cora Timken Burnett (t). **51 naturepl.com:** Wild Wonders of Europe / Sa (br). **SuperStock:** Dray van Beeck / NiS / Minden Pictures (ca). **52 Alamy Stock Photo:** BIOSPHOTO. **53 Bridgeman Images:** Granger (crb); Wolverhampton Art Gallery (tr). **54 Dreamstime.com:** Seadam. **55 naturepl.com:** Franco Banfi (crb); Joel Sartore / Photo Ark (tr); Alex Mustard (cra). **56-57 Getty Images / iStock:** pleshko74 (background). **naturepl.com:** Colin Monteath / Hedgehog House / Minden. **56 Science Photo Library:** ANN RONAN PICTURE LIBRARY / HERITAGE IMAGES (tr). **57 naturepl.com:** Alex Hyde (cla). **58-59 Bridgeman Images:** The Maas Gallery, London. **60-61 Alamy Stock Photo:** Aleksei Egorov (background). **naturepl.com:** Alex Mustard (t). **61 Bridgeman Images:** National Museums Scotland (tr). **Flickr.com:** jomilo75 (crb). **naturepl.com:** Joel Sartore / Photo Ark (br). **62-63 Getty Images / iStock:** Hein Nouwens (background). **62 Getty Images:** redhumv (c). **63 Getty Images / iStock:** Hein Nouwens (br). **Wikipedia:** National Museum Cardiff (tr). **64 Mark Gee / theartofnight** (tr). **65 Alamy Stock Photo:** Alina Melenteva (background). **Bridgeman Images:** Barnes Foundation (tr). **naturepl.com:** Ingo Arndt / Minden (br). **66-67 Getty Images:** ilbusca (background). **67 Bridgeman Images** (tc). **naturepl.com:** Ingo Arndt (cra); MYN / Carsten Krieger (cr); Joel Sartore / Photo Ark (br). **68-69 Getty Images / iStock:** Natalia Churzina (background). **69 Alamy Stock Photo:** CPA Media Pte Ltd (cra). **70-71 Getty Images / iStock:** Natalia Churzina (background). **70 Bridgeman Images:** NPL - DeA Picture Library (tl). **Wikipedia:** Sidney Hall (crb). **71 Alamy Stock Photo:** BIOSPHOTO (tr). **Dreamstime.com:** Boule13 (br); Bjorn Wylezich (bc). **naturepl.com:** MYN / Jerry Monkman (bl). **72 Science Photo Library:** Nuno Miguel Abreu Rodrigues / Amazing Aerial Agency. **73 naturepl.com:** Chris Gomersall / 2020VISION (bc). **Shutterstock.com:** aksol (background). **74-75 Getty Images / iStock:** ilbusca (background). **74 Alamy Stock Photo:** Penta Springs Limited. **75 Alamy Stock Photo:** World History Archive (cra). **Dreamstime.com:** Tetiana Kovalenko (bc). **76 naturepl.com:** Franco Banfi (tl). **76-77 Alamy Stock Photo:** Julian Parton (t). **Getty Images / iStock:** antiqueimgnet (background). **78-79 Shutterstock.com:** Evgeny Turaev (background). **78 Alamy Stock Photo:** Chronicle (tr). **naturepl.com:** Tui De Roy (bc). **79 Alamy Stock Photo:** Anan Kaewkhammul (tr); NAPA (cr); Michael S. Nolan (br). **Bridgeman Images** (l). **80 Getty Images:** Finnbarr Webster. **81 Alamy Stock Photo:** www.pqpictures.co.uk (crb). **Bridgeman Images:** Look and Learn (tr). **Getty Images / iStock:** MorePics (background). **82-83 Getty Images / iStock:** pleshko74 (background). **82 naturepl.com:** Sven Zacek. **83 Bridgeman Images:** Look and Learn. **84-85 Alamy Stock Photo:** Ingo Oeland. **Getty Images / iStock:** Monkographic (background). **85 naturepl.com:** Ole Jorgen Liodden (b). **86 naturepl.com:** Roland Seitre. **86-87 Getty Images / iStock:** Monkographic (background). **87 Bridgeman Images:** Christie's Images (tc). **naturepl.com:** David Fleetham (br). **88-89 Shutterstock.com:** alyaBigJoy (background). **88 Greg Lecoeur Underwater and Wildlife Photography**. **89 Alamy Stock Photo:** Photo 12 (bc). **Bridgeman Images:** Boltin Picture Library (cra). **90-91 Shutterstock.com:** alyaBigJoy (background). **90 Dreamstime.com:** Peter Leahy (br). **naturepl.com:** Franco Banfi (bl); Wild Wonders of Europe / Zankl (tl); Doug Perrine (bc). **91 Alamy Stock Photo:** CPA Media Pte Ltd. **92-93 naturepl.com:** Henley Spiers.

ACKNOWLEDGMENTS

94-95 Shutterstock.com: Alexander Sviridov (background). 94 Bridgeman Images: G. Dagli Orti / NPL - DeA Picture Library (br). Getty Images: Imagevixen (tc). 95 Bridgeman Images. 96-97 Shutterstock.com: Channarong Pherngjanda (background). 96 Alamy Stock Photo: Gueffier Franck (bc); Oriental Antiques UK - Asian Art Dealer and Appraiser (cra). 97 Bridgeman Images: Purix Verlag Volker Christen. 98 Alamy Stock Photo: The Picture Art Collection (cl). Dreamstime.com: Patrick Guenette (background); Thomas Lukassek (tr). 99 Bridgeman Images: Dundee Art Galleries and Museums (bl). naturepl.com: Nick Upton (tr). 100-101 Alamy Stock Photo: Marcus Harrison - botanicals (background). 100 Getty Images: Ignacio Palacios. 101 Alamy Stock Photo: Martin Fowler (bl). Bridgeman Images: Christie's Images / Mondrian / Holtzman Trust (tc). 102-103 Getty Images / iStock: ilbusca (background). 102 naturepl.com: Franco Banfi. 103 Dreamstime.com: Fascinadora (tc). naturepl.com: Shane Gross (b); Lewis Jefferies (cra); MYN / Sheri Mandel (cr). 104-105 Getty Images / iStock: ilbusca (background). 104 Alamy Stock Photo: Kertu Saarits. 105 Shutterstock.com: Circumnavigation (tc); LouieLea (cr). 106 Getty Images: Connect Images RM / Steve Woods Photography (t). 106-107 Getty Images / iStock: GlobalP. Getty Images: ilbusca (background). 107 Bridgeman Images: Florilegius (crb). Shutterstock.com: Eric Isselee (tl); Hugh Lansdown (cr). 108 naturepl.com: Anup Shah (cla). Shutterstock.com: Morphart Creation (background). 108-109 Alamy Stock Photo: JUAN CARLOS MUOZ. 110-111 Shutterstock.com: Morphart Creation (background). 110 Bridgeman Images: British Library archive (bc). Science Photo Library: JEAN-BERNARD NADEAU / LOOK AT SCIENCES (tl). 111 Alamy Stock Photo: LWM / NASA / LANDSAT (b). 112-113 Getty Images / iStock: Mateusz Atroszko (background). 112 Alamy Stock Photo: imageBROKER.com GmbH & Co. KG. 113 Alamy Stock Photo: ArteSub (crb); Charles Walker Collection (bl). BluePlanetArchive.com: Masa Ushioda (cr). Getty Images: Brown Bear / Windmill Books / Universal Images Group (tc). Shutterstock.com: Ivanenko Vladimir (br). 114 Bridgeman Images: David Lees Photography Archive. 115 naturepl.com: Marie Read (cra); Richard Robinson (bl). 116-117 Dreamstime.com: Roman Egorov (background). 116 Alamy Stock Photo: Andrey Nekrasov (b). 117 Alamy Stock Photo: Science History Images (tr). Getty Images: Art Wolfe (bc). 118-119 naturepl.com: Alex Mustard (t). Shutterstock.com: Bodor Tivadar (background).

118 Getty Images: Brent Durand (br). 119 naturepl.com: Brandon Cole (br). 120 Getty Images: Mark Newman. 121 Alamy Stock Photo: imageBROKER.com GmbH & Co. KG (tl); Alina Melenteva (background). Bridgeman Images: Sainsbury Centre for Visual Arts / Robert and Lisa Sainsbury Collection (crb). 122 Alamy Stock Photo: Edd Westmacott. (bc). 122-123 BluePlanetArchive.com: Richard Herrmann. 123 Alamy Stock Photo: The Reading Room (crb). 124-125 Getty Images: VCG. 126-127 Dreamstime.com: Patrick Guenette (background). 126 Bridgeman Images: Natural History Museum, London (t). 127 naturepl.com: Melvin Grey (br); Markus Varesvuo (tc); Markus Varesvuo (crb). 128 Getty Images: Nastasic (background). naturepl.com: Shane Gross (tl). 129 Brooke Pyke. 130-131 Dreamstime.com: Maria Komleva (background). Getty Images: Sean Weekly Photography (b). 131 Alamy Stock Photo: The Protected Art Archive (cla). 132-133 Getty Images: mikroman6 (background). 132 Getty Images: Bernard Radvaner (b). 133 Alamy Stock Photo: Lebrecht Music & Arts (tr). Shutterstock.com: Mr.wutthiphat vimuktanont (br). 134 Getty Images: Ernst Haas (tl). Shutterstock.com: Marzolino (bc). 134-135 Getty Images: mikroman6 (background). 135 Alamy Stock Photo: TTstudio. 136-137 naturepl.com: Tui De Roy (b). Shutterstock.com: Gwoeii (background). 136 Bridgeman Images: Universal History Archive / UIG (cra). 137 Alamy Stock Photo: The Natural History Museum (cra). 138-139 Alamy Stock Photo: Richard Bedford. 140 Alamy Stock Photo: Associated Press (tl). Depositphotos Inc: vectortatu (background). Shutterstock.com: Anton Rodionov (cr). 141 Alamy Stock Photo: Universal Images Group North America LLC. 142-143 Depositphotos Inc: vectortatu (background). 142 Getty Images: Peter Adams (t). 143 Getty Images: THEPALMER (tr). 144-145 Getty Images / iStock: VladisChern (background). 144 Sarah Crofts (t). 145 Alamy Stock Photo: blickwinkel (br). naturepl.com: Sylvain Cordier (tr, cra). 146-147 Getty Images: powerofforever (background). 146 Bridgeman Images: Lowe Art Museum / Gift of Drs. Ann and Robert Walzer (crb). Wikipedia: ESA / A.Gerst (tl). 147 Alamy Stock Photo: GRANGER - Historical Picture Archive (crb). 148-149 Alamy Stock Photo: CPA Media Pte Ltd (t). Shutterstock.com: alyaBigJoy (background). 149 naturepl.com: Guy Edwardes (crb). 150 Bridgeman Images: Bonhams, London, UK (clb); Fine Art Images (tl). 150-151 Shutterstock.com: alyaBigJoy (background). 151 Getty Images: Keystone / Stringer (b).

152-153 Getty Images: MIGUEL RIOPA / AFP. 154-155 Art Gallery of Greater Victoria: Stephen Topfer, Given in memory of Theo Wiggan by her Ikebana Students and Friends (t). Shutterstock.com: Marzufello (background). 155 Reuters: Mainichi Shimbun (br). 156-157 Shutterstock.com: Kuznetsova Darja (background). 156 Getty Images / iStock: USO (c). naturepl.com: Simon Bennett / BIA / Minden (cl); Jan Wegener / BIA / Minden (tl); Espen Bergersen (bl). 157 Getty Images / iStock: AntonioMasiello (br). 158-159 Alamy Stock Photo: Markus Mayer (background). 158 naturepl.com: Shane Gross (b). 159 Alamy Stock Photo: WaterFrame (tl). Bridgeman Images: Florilegius (bc). 160-161 Dreamstime.com: Ladadikart (background). 160 Bridgeman Images: Florilegius. 161 Alamy Stock Photo: Pix (cra). 162-163 Getty Images: powerofforever (background). 162 Alamy Stock Photo: Newscom (clb). Dreamstime.com: Andreyi Armiagov (cra). 163 Alamy Stock Photo: Shawshots. 164-165 Depositphotos Inc: vectortatu (background). 164 Alamy Stock Photo: Historic Collection (b). 165 Alamy Stock Photo: Heritage Image Partnership Ltd (crb). Shutterstock.com: DTM Media (tl). 166-167 Getty Images: Andrea_Hill (background). 166 Getty Images: Westend61 (b). 167 Bridgeman Images: Granger (tr); The Trustees of the British Museum: (bc). Getty Images: Whitworth Images (cr). naturepl.com: Sebastian Kennerknecht / Minden (br). 168-169 Valter Bernardeschi. 170-171 Getty Images / iStock: MorePics (background); Nicolas Tolstoï (t). 170 Science Photo Library: Science Source (tl). 172-173 Shutterstock.com: aksol (background). 172 Alamy Stock Photo: Zoom Historical. 173 Alamy Stock Photo: Danita Delimont Creative (tr). naturepl.com: Bryan and Cherry Alexander (bc). 174 Getty Images: sergei gladyshev. 175 Bridgeman Images: Look and Learn (tr). Getty Images: Culture Club (br). Shutterstock.com: Bodor Tivadar (background). 176 Alamy Stock Photo: Sam Kovak (br). 176-177 Shutterstock.com: Bodor Tivadar (background). 177 Vince Burton. 178-179 naturepl.com: Flip Nicklin / Minden (t). Shutterstock.com: Kseniakrop (background). 178 Bridgeman Images: Granger (crb). 180-181 Bridgeman Images: Photo Josse. Shutterstock.com: aksol (background). 181 Alamy Stock Photo: Marko Steffensen (bc). naturepl.com: Orsolya Haarberg (tr). 182 naturepl.com: Bryan and Cherry Alexander (tr). 182-183 naturepl.com: Tony Wu (b). Shutterstock.com: Arthur Balitskii (background). 183 Bridgeman Images: Museum of New Zealand te Papa Tongarewa (tc). 184-185 Alamy Stock Photo: Patrick Guenette (background). 184 Alamy Stock Photo: Niday Picture Library (tl). Shutterstock.com: David Osborn (bc). 185 naturepl.com: Jordi Chias (b). 186-187 Greg Lecoeur Underwater and Wildlife Photography. 188 Alamy Stock Photo: Vintage Travel and Advertising Archive (tr). 188-189 naturepl.com: Stefan Christmann (b). Shutterstock.com: alyaBigJoy (background). 190-191 Shutterstock.com: alyaBigJoy (background). 190 Joshua Vela-Fonseca (bl). 191 Getty Images: Florilegius (cra). Getty Images / iStock: Rixipix (br). naturepl.com: Klein & Hubert (bc); Yva Momatiuk & John Eastcott / Minden (bl). 192-193 Alamy Stock Photo: Artur Balytskyi (background). Greg Lecoeur Underwater and Wildlife Photography. 192 Alamy Stock Photo: The History Collection (cra). 193 Alamy Stock Photo: Penta Springs Limited (tc). 196-197 Getty Images / iStock: Hein Nouwens (background). 196 BluePlanetArchive.com: Ethan Daniels (b). 197 Bridgeman Images: Archives Charmet (crb). Science Photo Library: DAVID SALVATORI, VW PICS (tl). 198-199 Shutterstock.com: Sergey Kohl (background). 198 Alamy Stock Photo: S.E.A. Photo. 199 Bridgeman Images: AF Fotografie (tc). naturepl.com: Solvin Zankl (cr). Science Photo Library: Steve Gschmeissner (br); Alex Mustard / NPL (cra). 200-201 Alamy Stock Photo: Heritage Image Partnership Ltd. 202-203 Dreamstime.com: Patrick Guenette (background). 202 Dreamstime.com: Picture Partners (cr). naturepl.com: David Fleetham (bc); MYN / Sheri Mandel (tr). 203 Bridgeman Images: Prismatic Pictures. 204-205 Getty Images / iStock: Andrii-Oliinyk (background). naturepl.com: Alex Mustard. 205 Alamy Stock Photo: Nature Picture Library (tr). 206 naturepl.com: Jordi Chias (tl). 206-207 Getty Images / iStock: Andrii-Oliinyk. 207 Kahi Ching: on display at Mokupāpapa Discovery Center in Hilo (b). Science Photo Library: Richard J. Green (tr). 208-209 Shutterstock.com: Hein Nouwens (background). 208 Getty Images: ilbusca (tl). 209 Alamy Stock Photo: eFesenko (cb). Getty Images / iStock: Denja1 (br). naturepl.com: Alex Mustard (cr). Shutterstock.com: DiveIvanov (cra). 210-211 Dreamstime.com: Andrii Oliinyk (background). naturepl.com: Inaki Relanzon. 210 Getty Images: Bettmann (tl). 211 naturepl.com: Juergen Freund (br). 212-213 Dreamstime.com: Maart (background). naturepl.com: Alex Mustard. 213 Dreamstime.com: Boon Leng Teo (cra). OceanwideImages.com: Rudie Kuiter (br).

214-215 Dreamstime.com: Nataliia Golovanova (background). **214 Bridgeman Images:** CCI (bl); Museum of New Zealand te Papa Tongarewa / Donne Collection, Purchased 1905 (tr). **Dreamstime.com:** Isselee (cr). **John E Randall:** Bishop Museum, Honolulu (br). **Shutterstock.com:** AdrianNunez (cra). **215 naturepl.com:** Alex Mustard. **216-217 Getty Images / iStock:** ilbusca (background). **David Liittschwager**. **216 Bridgeman Images:** Photo Josse (clb). **217 naturepl.com:** Pete Oxford / Minden (br). **Science Photo Library:** Angel Fitor (cra). **Shutterstock.com:** Benny Marty (cr). **218 Bridgeman Images:** Archives Charmet. **218-219 Getty Images / iStock:** ilbusca (background). **219 Bridgeman Images:** Natural History Museum, London (crb). **naturepl.com:** Richard Herrmann / Minden (tr). **220-221 Shutterstock.com:** vector_ann (background). **220 naturepl.com:** David Fleetham (b). **221 123RF.com:** imagesource (cra). **Bridgeman Images:** Patrice Cartier (cla); NPL - DeA Picture Library (bc). **Dreamstime.com:** Caan2gobelow (crb); Nialldunne24 (cr). **222-223 Dreamstime.com:** Jeneses Imre (background). **222 Bridgeman Images:** Museum of Fine Arts, Houston / Museum purchase funded by the Shell Oil Company Foundation at "One Great Night in November" (crb). **naturepl.com:** Tony Wu (t). **223 Bridgeman Images:** Infatti (tr). **224-225 naturepl.com:** Tony Wu. **225 Alamy Stock Photo:** History of America (bc). **Shutterstock.com:** alyaBigJoy (background). **226-227 Shutterstock.com:** alyaBigJoy (background). **226 Dreamstime.com:** Jemma Craig (bl). **Getty Images / iStock:** elmvilla (tl). **naturepl.com:** Alex Mustard (clb). **Shutterstock.com:** Andrew Sutton (cl). **227 Alamy Stock Photo:** CPA Media Pte Ltd (b). **228-229 Clayton Harris**. **230 Getty Images / iStock:** AskinTulayOver (tl). **230-231 Bridgeman Images:** Sotheby's. **Shutterstock.com:** Very_Very (background). **232-233 Shutterstock.com:** Very_Very (background). **232 Bridgeman Images:** Christie's Images (ca). **233 Bridgeman Images:** Christie's Images (tr). **Getty Images:** Nick Brundle Photography (b). **234-235 Getty Images / iStock:** Nastasic (background). **234 naturepl.com:** Juergen Freund. **235 Dreamstime.com:** Zeng Hu (bc). **naturepl.com:** Georgette Douwma (tr, cr, crb). **236 Alamy Stock Photo:** Aleh Varanishcha. **237 Bridgeman Images:** Luisa Ricciarini (bc). **naturepl.com:** SCOTLAND: The Big Picture (tc). **238 Getty Images:** ANDREAS SOLARO / AFP. **238-239 Shutterstock.com:** Christos Georghiou (background). **239 Alamy Stock Photo:** UPI (crb). **Wikipedia:** Alphonse de Neuville / Henri Théophile Hildibrand (tr). **240-241 Alamy Stock Photo:** Benny Marty. **Dreamstime.com:** Channarong Pherngjanda (background). **240 Wikipedia:** Musée du quai Branly (crb). **241 Bridgeman Images. 242-243 Dreamstime.com:** Channarong Pherngjanda (background). **242 Alamy Stock Photo:** Marko Steffensen (br). **Dreamstime.com:** Diverstef (c). **Shutterstock.com:** Rich Carey (bc); Martin Prochazkacz (bl). **244-245 Shutterstock.com:** Nata_Alhontess (background). **244 naturepl.com:** Alex Mustard. **245 Alamy Stock Photo:** Blue Planet Archive (bc). **Etsy, Inc.:** BeautyandBeasty (tr). **naturepl.com:** Andy Murch (br). **Shutterstock.com:** Rich Carey (bl). **246-247 BluePlanetArchive.com:** Mark Strickland. **248-249 Shutterstock.com:** Kseniakrop (background). **248 Bridgeman Images / © DACS 2024** (tl). **249 Fabi Fregonesi** (t). **naturepl.com:** Ingo Arndt (bc). **250-251 Getty Images / iStock:** Evgeniy Zotov (background). **250 Bridgeman Images** (bl). **naturepl.com:** David Shale (tr). **251 Science Photo Library:** Dante Fenolio. **252 Alamy Stock Photo:** Penta Springs Limited. **253 Alamy Stock Photo:** The Natural History Museum (tr). **BluePlanetArchive.com:** e-Photo / Hideyuki Utsunomiya (b). **254-255 Shutterstock.com:** roughedges_stock. **254 Alamy Stock Photo:** PF-(usna1). **255 naturepl.com:** Constantinos Petrinos (cra); Paul D Stewart (crb). **256-257 Shutterstock.com:** roughedges_stock (background). **256 David Liittschwager** (ca). **257 Bridgeman Images:** John Bethell (tr). **Dorling Kindersley:** Frank Greenaway / Weymouth Sea Life Centre (bc). **naturepl.com:** Alex Mustard (bl); David Shale (br). **258 Dorling Kindersley:** Gary Ombler / University of Aberdeen (tc). **258-259 Alamy Stock Photo:** GRANGER - Historical Picture Archive (t). **Shutterstock.com:** Very_Very (background). **260-261 Shutterstock.com:** Very_Very (background). **260 Alamy Stock Photo:** Leopold von Ungern (t). **The Trustees of the British Museum** (clb). **261 Alamy Stock Photo:** Old Books Images (tr). **262-263 Shutterstock.com:** Morphart Creation (background). **262 Sumer Verna**. **263 Alamy Stock Photo:** Iconographic Archive (crb). **Getty Images:** Thomas Bannenberg / 500px (cr); Thomas Kurmeier (cra). **naturepl.com:** Andy Murch (tc). **264-265 Dreamstime.com:** Patrick Guenette (background). **264 Alamy Stock Photo:** BIOSPHOTO (ca). **naturepl.com:** Pete Oxford / Minden (bl). **Science Photo Library:** Dr George Gornacz (cl). **265 Alamy Stock Photo:** Florilegius (t). **266-267 Dreamstime.com:** Patrick Guenette (background). **266 Alamy Stock Photo:** World History Archive (tl). **Bridgeman Images:** A. Dagli Orti / NPL - DeA Picture Library (cb). **267 naturepl.com:** Alex Mustard. **268-269 BluePlanetArchive.com:** Doug Perrine. **Getty Images:** powerofforever (background). **269 Alamy Stock Photo:** Connect Images (cra); WaterFrame (tr); SOTK2011 (br). **270-271 Alamy Stock Photo:** Zoonar GmbH (t). **Getty Images / iStock:** ilbusca (background). **270 BluePlanetArchive.com:** Ethan Daniels (cb). **272-273 Getty Images / iStock:** ilbusca (background). **272 Bridgeman Images:** Superstock (bc). **naturepl.com:** Philippe Clement (tr); Joel Sartore / Photo Ark (cla). **Shutterstock.com:** Asinka Photography (tl); Kasia (clb). **273 Bridgeman Images:** Iberfoto. **274-275 Getty Images:** ilbusca (background). **274 Alamy Stock Photo:** WaterFrame (cla). **Bridgeman Images:** DaTo Images (c). **naturepl.com:** Brandon Cole (clb); David Fleetham (br). **275 Bridgeman Images:** A. De Gregorio / NPL - DeA Picture Library (tr). **naturepl.com:** Henley Spiers (br). **276-277 Tanya Houppermans**. **278-279 Alamy Stock Photo:** Taisiya Geraskina (background). **278 Getty Images:** Wild Horizon. **279 Alamy Stock Photo:** FOODSTUFF (cr); Science History Images (tc); Penta Springs Limited (bc). **Getty Images:** Christian Zappel (crb). **naturepl.com:** Tony Wu (br). **280 BluePlanetArchive.com:** Reinhard Dirscherl (tl). **Bridgeman Images:** Royal Asiatic Society (cr). **281 Getty Images:** Arne Hodalic / CORBIS. **282 Alamy Stock Photo:** Jakob Ziegler (br). **David L. Mearns / Blue Water Recoveries Ltd:** (bl). **naturepl.com:** Juergen Freund (tl). **283 Getty Images:** Universal History Archive. **284-285 Shutterstock.com:** Marzufello (background). **284 naturepl.com:** Norbert Wu / Minden (tl). **285 David Liittschwager** (b). **naturepl.com:** Suzi Eszterhas / Minden (tr). **286-287 Dreamstime.com:** Alisa Aleksandrova (background). **286 Getty Images:** Petar Belobrajdic / 500px. **287 Science Photo Library:** Dante Fenolio (tr). **288-289 Getty Images / iStock:** ilbusca (background). **288 naturepl.com:** David Hall. **289 California Academy of Sciences:** Douglas J. Long (crb). **NOAA:** (br). **Science Photo Library:** MMA / PURCHASE / SCIENCE SOURCE (tr). **290-291 Adobe Stock:** Alisles (background). **290 Bridgeman Images:** Sainsbury Centre for Visual Arts / Robert and Lisa Sainsbury Collection (crb); The Stapleton Collection (t). **291 naturepl.com:** Roberto Rinaldi (cr); Kim Taylor (cra). **Science Photo Library:** Alexander Semenov (bl). **Shutterstock.com:** Warren Metcalf (br). **2**92-293 **Shutterstock.com:** Kseniakrop (background). **292 naturepl.com:** Doc White. **293 Bridgeman Images** (tr). **Getty Images:** Werner Forman (bc). **294-295 Shutterstock.com:** Kseniakrop (background). **294 Science Photo Library:** B. Murton / Southampton Oceanography Centre (tr). **295 University of Washington:** D.S. Kelley and M. Elend, University of Washington; URI ROV Hercules; NOAA Ocean Exploration. **296 Alamy Stock Photo:** Blue Planet Archive (b); The Natural History Museum (cra). **297 Alamy Stock Photo:** PF-(usna1) (cb). **Dreamstime.com:** Maart (background). **MBARI** (ca)(cra)(tl). **298-299 Getty Images / iStock:** ilbusca (background). **298 NOAA** (tr). **UWA Deep Sea Research Centre:** Minderoo-University of Western Australia (UWA) Deep Sea Research Centre and the Tokyo University of Marine Science and Technology (bc). **299 Alamy Stock Photo:** Album. **300-301 Alamy Stock Photo:** IanDagnall Computing. **300 Alamy Stock Photo:** Xinhua (tr). **301 Bridgeman Images:** Look and Learn